CW01649814

Higher

Human Biology

2000 Exam

2001 Exam

2002 Exam

2003 Exam

2004 Exam

Leckie × Leckie

First exam published in 2000.
Published by Leckie & Leckie, 8 Whitehill Terrace, St. Andrews, Scotland KY16 8RN tel: 01334 475656 fax: 01334 477392
enquiries@leckieandleckie.co.uk www.leckieandleckie.co.uk

Leckie & Leckie Project Team: Andrea Collington; Peter Dennis; Bruce Ryan

ISBN 1-84372-243-7

A CIP Catalogue record for this book is available from the British Library.

Printed in Scotland by Scotprint.

Leckie & Leckie is a division of Granada Learning Limited, part of ITV plc.

Acknowledgements

Leckie & Leckie is grateful to the copyright holders, as credited, for permission to use their material. Every effort has been made to trace the copyright holders and to obtain their permission for the use of copyright material. Leckie & Leckie will gladly receive information enabling them to rectify any error or omission in subsequent editions.

The Scotsman for two bar graphs (2002 p 8);

The following companies have very generously given permission to reproduce their copyright material free of charge:

HarperCollins for the diagram taken from Advanced Human Biology by J. Simpkins and J.I. Williams (2003 p 14).

Pocket answer section for SQA Higher Human Biology 2000 – 2004

Published by Leckie & Leckie Ltd, 8 Whitehill Terrace, St Andrews, Scotland, KY16 8RN
tel: 01334 475656, fax: 01334 477392, enquiries@leckieandleckie.co.uk, www.leckieandleckie.co.uk

Higher Human Biology 2000

Section A

1.	B	**11.**	D	**21.**	D
2.	A	**12.**	D	**22.**	B
3.	C	**13.**	B	**23.**	B
4.	D	**14.**	B	**24.**	A
5.	D	**15.**	C	**25.**	C
6.	B	**16.**	B	**26.**	A
7.	C	**17.**	D	**27.**	B
8.	A	**18.**	A	**28.**	D
9.	B	**19.**	C	**29.**	A
10.	A	**20.**	C	**30.**	A

Section B

- *(brackets) indicate useful but not essential*
- *solidus / indicates alternative*
- *<u>underline</u> indicates essential word phrase/idea required*
- *spelling should be at least phonetic to gain mark, except where specified otherwise eg 1(a)*

1. (*a*) **1** uracil
2 cytosine
3 thymine
(*correct spelling only*)

(*b*) **1** ribose (sugar)
2 phosphate/Pi (group)

(*c*) ribosome

(*d*) to carry code for manufacture/synthesis of protein/enzyme (1)
<u>enzymes</u> are essential/catalysts for cell metabolism (1)

2. (*a*) **F**(T)
(F)**F**
FT

(*b*) (i) pyruvic acid/pyruvate (*correct spelling only*)

(ii) oxygen

(iii) because less ATP/energy is produced or because waste products contain much energy or because breakdown of glucose is incomplete

(iv) glycogen (*correct spelling only*)

3. (*a*) (i) protein

(ii) nucleic acid changed/damaged/altered/absent **or** virus attenuated

(*b*) (i) viral DNA/RNA enters cell; viral DNA replicates; viral protein is manufactured

(ii) the cell (membrane) bursts/ruptures **or** lysis takes place

3. (*c*) because antibodies are specific **or** antibodies only recognise one type of antigen/virus **or** because viruses have different receptor/attachment sites

(*d*) (i) 350–650

(ii) 1975–76, 1977–78 and 1981–82 **or** 1976, 1978, 1982

(iii) because cilia cannot remove viruses/microbes from lungs (*not bacteria, dust, germs*)

(iv) the population is immune from previous exposure to this virus **or** there is a different weather pattern **or** a vaccination programme was introduced **or** there were different strains/types of viruses

4. (*a*) (i) B

(ii) change of order of bases **or** removal/deletion/addition/insertion/inversion/substitution of bases

(*b*) 50%

(*c*) pituitary (gland)

(*d*) (i) Down's syndrome

(ii) haemophilia

5. (*a*) (i) seminiferous tubules

(ii) activation/nutrition of sperm/provides energy/fluid medium/stimulates muscular contraction of female tract

(*b*) (i) *a cross (X) on sperm duct between testis and label-line to prostate gland*

(ii) because testosterone is carried in the blood

6. (*a*) *vertical line of any length at second week – between second and third vertical line*

(*b*) level → steep/sharp rise → steady rise
(*3 distinct stages described – 2 marks*)
(*2 distinct stages plus correct numerical reference – 2 marks*) (*1 stage plus 1 correct numerical reference – 1 mark*)

(*c*) (i) corpus luteum/ovary

(ii) placenta

(*d*) to stimulate/start/cause milk production/lactation (*not: to produce milk*)

Higher Human Biology 2000 (cont.)

7. (*a*) to remove excess tissue fluid/to carry lymph

(*b*) valves are present, which prevent the backflow of lymph (1)

the flow is maintained by muscular action/body movement (1)

blood pressure of tissue fluid is higher than that of lymph (1) (*any two*)

(*c*) because of an increase in numbers/activity of white blood cells/macrophages/ lymphocytes (1)
or because of accumulation of fluid **or** because of inflammation

(*d*) it carries out phagocytosis **or** it removes/digests foreign particles/engulfs bacteria (1)

(*e*) by diffusion **or** *description of diffusion*

(*f*) because fluid is lost from the blood **or** because of high surface area/cross section area of capillaries

or because of increased friction/resistance to flow

8. (*a*) (i) 1 : 3

(ii) directly proportional/as one rises, so does the other/as one falls, so does the other

(iii) 36·7 °C
(*accept as low as 36·68 °C*)

(iv) because changes to sweat production follow changes to skin temperature

(*b*) vasodilation **or** relaxation of hair muscles **or** decreased metabolic rate

(*c*) hypothalamus

9. (*a*) A Somatic B Central/CNS
C Brain D Sympathetic

(*b*) (i) working in opposition/having opposite effects

(ii) parasympathetic stimulates digestive system whereas sympathetic system inhibits digestive system (1) eg peristalsis/blood flow stimulated by parasympathetic (1)

10. (*a*) retina

(*b*) (i) converging/sensory

(ii) synapse/synaptic cleft

(iii) impulses from a number of cells come together/ accumulate/are added (1) this is more likely to reach a threshold/cause impulse to cross gap (1)

(*c*) visual/cerebral cortex/cerebrum

11. (*a*) ten/many students in each group

(*b*) because they were organised into related groups/categorised

(*c*) (i) *One mark for both scales and one mark for plotting points. Line of best fit,* **or** *straight lines joining points are acceptable. Deduct one mark for: plotting to zero* **or** *less than 50% of graph paper used* **or** *x/y scales transposed.*

(ii) because early words can be rehearsed **or** transferred to LTM

(iii) because words remembered late are not displaced from STM

(iv) serial position effect

(*d*) **Experiment 1** words are only shown for 30s/video.

Experiment 2 words are read without pause.

12. (*a*) respiratory diseases

(*b*) (i) infectious and parasitic diseases

(ii) lack of immunisation programme/poor water supply/poor hygiene/poor medical care/overcrowding in developing countries
or vice versa

(*c*) 10 million

(*d*) infections and parasitic diseases

(*e*) (i) cancers and/or circulatory diseases

(ii) The children have not lived long enough to allow these diseases to develop

or these are diseases of old age **or** children have not been exposed to smoke/bad diet.

(*f*) Most children are breast fed.

(*g*) Lack of access to safe drinking water.

(*h*) because the developing countries account for a much larger percentage/proportion of the world population

Section C
solidus / indicates alternatives
one mark per line unless stated otherwise

1. A Circulation of blood

(i) Vena cava brings blood to the heart from body to right atrium

Right atrium pumps blood into right ventricle through tricuspid valve*

Right ventricle pumps blood to lungs via pulmonary artery

(ii) Pulmonary vein brings blood to the heart from the lungs to left atrium

Left atrium pumps blood to left ventricle through bicuspid valve*

Left ventricle pumps blood to body via aorta

*Atrio-ventricular (AV) valve gains one mark if other two names not used

In addition, in either (i) or (ii)

valves prevent back flow

right ventricle has lower pressure/weaker than left ventricle

contraction of heart muscle is "systole"/relaxation is "diastole"

arteries lead to arterioles/capillaries **or** capillaries lead to veins/venules

right side of heart deals with deoxygenated blood and/**or** *vice versa*

B. Filtration and reabsorption in the kidney

(i) Glomerulus is a knot/bundle of capillaries

Blood (plasma) is filtered from the glomerulus to Bowman's capsule

Red blood cells *and/or* proteins are too large to pass through filter

Water/glucose/amino acids/vitamins/salts/minerals/urea pass through filter (*any three*)

High pressure due to different cross section of arterioles

High surface area for easy/quick filtration **or** mention of high filtration volume (175 litres/24h)

(ii) Useful materials which have been filtered are reabsorbed

eg much water/salts and all glucose/amino acids, but not urea (*any three*)

(most) Reabsorption takes place in proximal convoluted tubule by active transport/ requires expenditure of energy

1. B (ii) **continued**

Water reabsorbed/salts removed in Loop of Henle

ADH controls reabsorption of water

ADH acts in collecting duct

ADH makes tubules more permeable

ADH produced when there is a shortage of water

2. A. Role of Lipids in the Body (maximum – 8 marks)

Energy store

Useful because insoluble **or** 2× more energy than carbohydrate weight for weight

Heat insulation

Protection from physical damage eg fat pads of feet/hands

Transport of certain vitamins (A,D,E,K)

Component of cell membranes

Diagram to show phospholipid bilayer **or** describe phospholipids in two layers

Some hormones are lipids/steroids eg testosterone/oestrogen/progesterone

Insulation of nerve fibres

By myelin sheath

Sebum/wax waterproofing of skin/keeps skin supple/protection from bacteria

B. Disruption of the Carbon Cycle (maximum – 8 marks)

Plants gain carbon by absorbing CO_2 from the atmosphere during photosynthesis

Animals gain carbon by eating plants or other animals

Carbon is returned to the atmosphere as CO_2 through respiration in animals and plants

CO_2 is produced as result of increased human activity/increased population/industrialisation eg burning fossil fuels (coal, oil and/or gas)/ biomass/trees (*any two*)

Removal of plants/forests/deforestation (to be replaced by buildings/agricultural land)

This is causing an increase in CO_2 concentrations in the atmosphere

This in turn is likely to be causing global warming/"greenhouse effect"

CO_2 acts as a blanket, retaining sun's heat

Excess methane from paddy fields/cattle has a similar effect

This will cause icecaps to melt/water to expand and sea level to rise

Results in major changes to weather patterns/drought/flooding/storms (*any two*)

Higher Human Biology 2001

Section A

1.	D	11.	C	21.	A
2.	B	12.	D	22.	B
3.	C	13.	A	23.	C
4.	C	14.	B	24.	A
5.	B	15.	D	25.	C
6.	A	16.	D	26.	D
7.	C	17.	A	27.	B
8.	B	18.	C	28.	B
9.	D	19.	B	29.	C
10.	A	20.	D	30.	B

Section B

solidus / indicates alternatives
(brackets) indicate desirable but not essential

1. (*a*) (i) active site
(ii) mitochondrion
(iii) water

(*b*) It changes the shape of the active site.
or It attaches to/blocks the active site.

(*c*) They are activators/co-factors/co-enzymes.

(*d*) (i) All active sites are occupied.
or Enzyme concentration is limiting factor.

(ii) Add more enzyme.
or Increase the concentration of the enzyme.

2. (*a*) (i) (1) sodium 1 6
(2) potassium 30 1

(ii) by active transport

(iii) decrease
Glucose is a source of energy (for making ATP/for respiration).
ATP/energy needed for active transport.

(*b*) A Cells burst because water enters (by osmosis).
B Cells stay same because solution is isotonic/same concentration.
C Cells shrink because water lost (by osmosis).

3. (*a*)

Blood group	*Antigens present on surface of red blood cells*	*Antibodies present in plasma*
A	A	**anti-B**
B	**B**	anti-A
AB	**A and B**	**none**
O	**none**	**anti-A and anti-B**

(*b*) A and O

(*c*) (i) female AB male AO

3. (*c*) (ii)

male gametes \ female gametes	**A**	**B**
A	**AA**	**AB**
O	**AO**	**BO**

(iii) 50

4. (*a*) (i) placenta
(ii) progesterone/oestrogen

(*b*) (i) cervix
(ii) thinner/less sticky/more watery

(*c*) (i) length/size of chromosomes/position of centromere/shape of chromosomes
banding patterns

(ii) karyo/type/-graph

(iii) by checking sex chromosomes (XY male and XX female)

(iv) whether there is an abnormal number of chromosomes
or whether fetus has Down's syndrome/Kleinfelter's etc

5. (*a*) (i) glomerulus
(ii) The blood vessel entering is wider than that leaving.
(iii) proximal (convoluted) tubule

(*b*) (i) reabsorption
(ii) 40
(iii) The urine would contain glucose/more water.

6. (*a*) (i) autonomic (nervous system)
(ii) medulla (oblongata)

(*b*) (i) SAN/sino-atrial node/ pacemaker
(ii) slows/decreases heart rate

(*c*) slows breathing/stimulates peristalsis/constricts pupil/decreases blood pressure/etc

7. (*a*) (i) Pulse rate 86 bpm
Stroke volume 130 cm^3

(ii) 110

(iii) 16·5 litres

(iv) As one increases so does other.
Stroke volume unchanged at high level of exercise.

(*b*) (i) 6
(ii) number of breaths per minute/breathing rate

7. (*c*) (i) 23 or 24

(ii)

Statement	*Tick* (✓)
The rate at which pulse rate changes is highest at low rates of oxygen uptake.	✓
When ventilation rate doubles, the rate of oxygen uptake doubles.	

(*d*) pulmonary artery

8. (*a*) Type of neurone – motor
Reason for answer
cell body at one end of neurone/ many dendrites present

(*b*) Name – **nucleolus** (*spelling must be correct*)
Function
manufacture of RNA/ribosomes/ ribosomal RNA

(*c*) (i) vesicle

(ii) neurotransmitter/ acetylcholine/noradrenalin

(*d*) (i) slows down/hinders/stops transmission of (nerve/electrical) impulse

(ii) autoimmune

(*e*) *arrow(s) must point towards nerve cell body*

(*f*) Impulse is transmitted to several points/destinations at same time.
This permits muscles/fingers to work in a coordinated way.

9. (*a*) Malnutrition: diet lacks vitamins/ minerals/sufficient protein/fibre/ sodium/calcium
Starvation: diet supplies sufficient energy (from fat)

(*b*) Natural ecosystems
Results in deforestation/erosion due to overgrazing
Local water supplies
results in reduced rainfall OR flooding OR silting of rivers OR nitrate/sewage pollution

(*c*) energy is lost at each link in food chain so the extra link/step in food chain results in energy loss

(*d*) increased yield
improved disease/frost/drought/resistance **or** improved storage/flavour etc

(*e*) (i) A chemical which kills pests/unwanted organisms.

(ii) It increases crop yield (because crop is less likely to be eaten/ damaged by pests).

(*f*) quicker/more precise/allows transfer of genes between different species

10. (*a*)

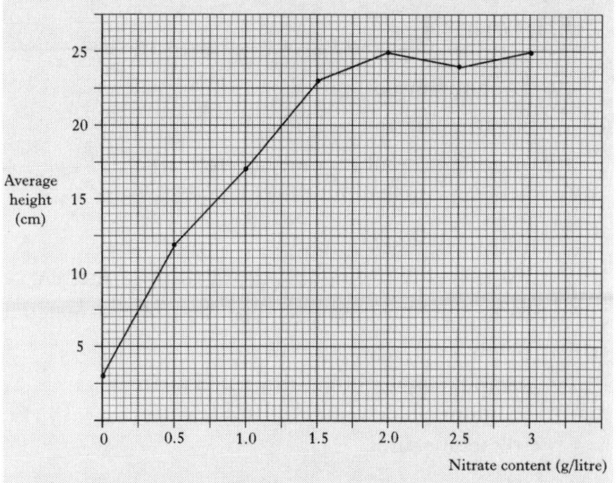

Correct y-axis with numbers and units (1)
Correct x-axis with numbers and units (1)
Correct plot by straight lines or line of best fit (1)
One mark reductions for:
Use of less than half of graph paper. ie half-size scale
Transposition of axes

(*b*) temperature/pH/volume of solution/type of seed/light intensity/air supply

(*c*) to provide oxygen (for roots)

(*d*) the experiment was repeated (ten times)

(*e*) number/surface area/colour of leaves **or** length of roots/(dry) weight

(*f*) Prediction
The plants would grow better/as well as others (at low nitrate concentrations).
Reason
They can fix/obtain their own nitrogen.

(*g*) It indicates the minimum amount of fertiliser required for optimum growth.

Section C

1 mark for each line unless specified otherwise
(brackets) = no marks
solidus / = alternative

1. A.

(i) **Encoding**

The means by which information is entered into memory eg sound, smell, taste, visual, semantic/meaning.

(Encoding is enhanced by) rehearsal, organisation, elaboration + description or examples.

Mnemonic devices + description or example.

(ii) **Storage**

LTM can store almost unlimited number of items.

STM has limited capacity of around 7 items/ called memory span.

More information can be held if it is chunked + example of chunking.

Excess information in STM will be displaced/

Higher Human Biology 2001 (cont.)

1. A (ii) **continued**

lost/transferred to LTM.

Localisation of memory in limbic system/cerebrum/hippocampus.

(iii) **Retrieval**

Retrieval is getting information out of memory.

it is aided by contextual cues eg sights, smells, sounds

Serial position effect described eg items at beginning and end of a list are easier remembered.

B.

(i) **Artificial active immunity**

Immunisation is the introduction of weakened/harmless form of virus/bacterium to the body.

Nucleic acid/cell contents damaged/attenuated, so will not cause the disease.

However, surface antigens are same, so will be recognised by body defences.

B-lymphocytes manufacture antibodies.

T-lymphocytes attack cells directly.

Memory cells are produced.

So, if same pathogen invades body again, there will be faster/stronger response.

Body now has (long-term) immunity to the disease.

One example of disease for which we are routinely vaccinated. eg polio, tetanus, diphtheria, whooping cough, meningitis, measles, mumps, rubella, TB.

(ii) **Artificial passive immunity**

Injection/inoculation of (ready-made) antibodies into the body.

A short lived effect.

because it does not induce an immune response from the body.

Allows body time to develop its own antibodies.

One example of disease for which this type of inoculation is used eg tetanus, rabies, hepatitis, measles.

(iii) **The impact of vaccination on childhood diseases**

Public inoculation/vaccination programmes have reduced death rates smallpox eradicated worldwide.

Developing countries still have high mortality due to common childhood diseases.

Examples of disease eg diphtheria, whooping cough, measles, tetanus, polio and TB.

Examples of diseases in each of three sections above must be different and separate to gain marks.

2. A.

Describe the biological basis of contraception.

Contraception is prevention of fertilisation after intercourse/pregnancy.

Fertile period lasts for a few days around day 14/mid point of cycle.

Rhythm methods of contraception rely on detection of this fertile period.

Can be detected by change in body temperature or changes in cervical mucus/mucus becomes thinner.

Hormonal contraceptives can be pills/injections/implants/morning after which contain synthetic oestrogen/progesterone.

Pills usually taken for 3 weeks/one pill taken each day.

Concentration of hormones in blood is artificially increased.

Causes negative feedback effect on pituitary gland.

Reduced production of FSH prevents maturation of ova/eggs.

Reduced production of LH prevents ovulation.

Prolonged/regular suckling acts as a contraceptive.

Explanation/labelled diagram of menstrual cycle. (1)

B.

Outline involuntary mechanisms of temperature control.

Hypothalamus detects/controls changes in blood temperature.

Also receives information from (thermo) receptors in the skin.

(responses to overheating)

Increased blood flow to skin causes increased heat loss (by radiation/convection)/vasodilation.

(Increased) sweat production which results in heat loss by evaporation.

Reduced metabolic rate resulting in less heat production.

(responses to cooling)

Reduced blood flow to skin reduces heat loss (by radiation/convection)/vasoconstriction.

Contraction of hair (erector pili) muscles make hair stand on end and trap a layer of air which insulates skin/prevents heat loss.

Increased metabolic rate causes increased heat production.

Increased release of adrenaline/thyroxine.

Shivering increases heat production.

Shivering is spasmodic contraction of muscles.

Higher Human Biology 2002

Section A

1.	C	**11.**	D	**21.**	C
2.	A	**12.**	B	**22.**	D
3.	C	**13.**	D	**23.**	A
4.	D	**14.**	A	**24.**	C
5.	C	**15.**	B	**25.**	A
6.	B	**16.**	C	**26.**	C
7.	B	**17.**	A	**27.**	B
8.	A	**18.**	C	**28.**	A
9.	B	**19.**	A	**29.**	C
10.	D	**20.**	D	**30.**	B

Section B

1. (*a*) (i) ribosome

(ii) protein

(iii) *any example of protein* eg enzyme, antibody, hormone

(*b*) (i) (rough) endoplasmic reticulum/ER

(ii) Example Golgi body/apparatus

Function packaging, processing of materials/proteins

(*c*) To allow large/insoluble molecules/mRNA to pass through membrane

2. (*a*) Bond **X** peptide bond

Molecule **Y** mRNA

(*b*) anti-codon

(*c*) *iso* → **arg** → **tyr** → **ala** → **leu**

(*d*) TGC

(*e*) $\frac{2}{3}$s or decimal equivalent (0·6 to 0·7 acceptable)

3. (*a*) (i) P **rr** T **Rr**

(ii) 6

(iii) *cross through white square*

(iv) No

There is not enough information to be certain whether Female S is homozygous or heterozygous.

(*b*) (i) polygenic

(ii) **height** **skin colour**

4. (*a*) (i) **A → D → B → C**

(ii) A

(iii) They result in variation in individuals or gametes/shuffling of alleles/genes

(*b*) (i)

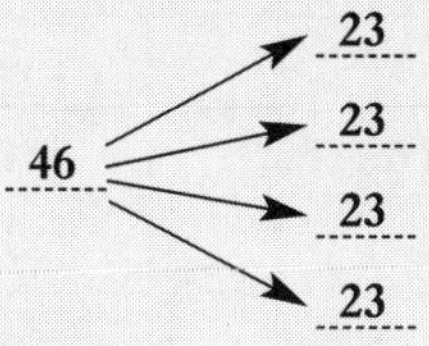

(ii) 2

(iii) seminiferous tubules/in sperm mother cells

5. (*a*) progesterone

(*b*) causes/stimulates ovulation

(*c*) **22–26 days**

(*d*) Difference It would not rise at the end of the cycle/It would stay low

Explanation because pituitary gland (which produces FSH) is inhibited (by progesterone)

6. (*a*)

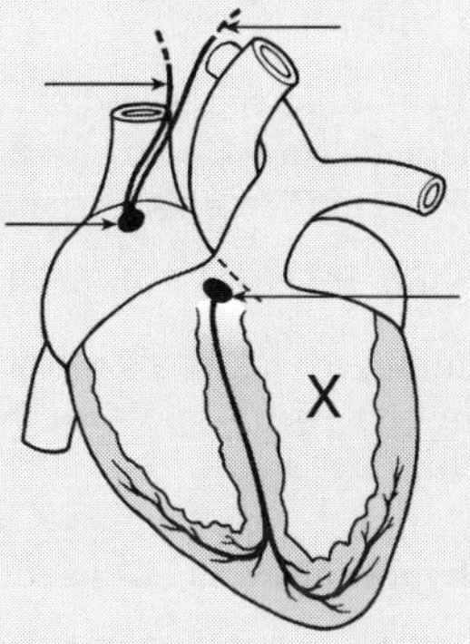

(*b*) They slow the heart (rate)

(*c*) (i) AVN (atrio-ventricular node)

(ii) It co-ordinates/stimulates the contraction of the ventricles

or

Sends impulses down Bundle of His

(iii) 0·8 s

7. (*a*) (i) lacteal

(ii) It removes/absorbs/carries fats/fatty acids + glycerol/lipids away from villus/gut

(*b*) (i) intrinsic factor

(ii) red blood corpuscle/cell

(*c*) hepatic portal vein

Higher Human Biology 2002 (cont.)

8. (*a*)

Substance	*Change*
Oxygen	**decrease**
Carbon dioxide	**increase**
Insulin	**increase**

(*b*) (i) It lowers glucose concentration of the blood
or
It stimulates conversion of glucose to glycogen

(ii) glucagon

9. (*a*) (i) People eat less in 1998 than in 1988

(ii) The figures for Scottish consumption in 1988

(iii) 1 The number/sex/class/type/age/
2 income group of people surveyed
Any ***two***

(*b*) 10·3%

(*c*) 126 g/person/week

(*d*) 1 Less milk, cream and cheese consumed/less consumption of foods high in fat/less consumption of sugar and preserves

2 More fruit and fruit products consumed

(*e*) planning school meals subsidies/content/farm quotas/subsidies/health advertising/health care

10. (*a*) A **hypothalamus** B **medulla (oblongata)**

(*b*) (i) motor/sensory/auditory/visual etc

(ii) It increases the surface area to allow for more nerve cells/neurones

(iii) It transfers information from/or links one hemisphere to the other

(*c*) A disorder particularly associated with memory loss is **Alzheimer's** disease. This disorder is due to the disappearance of cells which produce the neurotransmitter **acetylcholine** in the **limbic** system of the brain.

(*d*) Decrease in water concentration detected by hypothalamus/which stimulates increased secretion of ADH/which makes kidney tubules more permeable/so more water is removed from urine/returned to blood
Any ***three***

11. (*a*) 7·7s

(*b*)

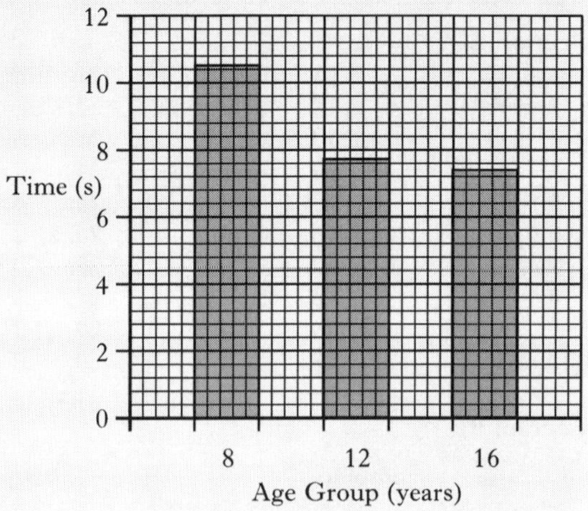

(*c*) (i) Because the numbers of groups of pupils tested are not the same/Because the numbers of trials are not the same

(ii) The sex distribution of the pupils
The method of obtaining fastest times was applied equally
That each group of pupils was dealt with in the same way
That the same maze was used each time

(*d*) social facilitation

12. (*a*)

Curve	*Year*
X	**1800**
Y	**1960**

Reason The death rate of young people was very high in 1800/or converse/more people survive to old age in 1960.

(*b*) better hygiene, better medical care, better diet, better housing, vaccination programme, antibiotics, etc in 1960
Any ***two***

(*c*) **X** 40 % **Y** 88 %

(*d*) **X** 0 to 10 years **Y** 80 to 90 years

13. (*a*) Both rise approx. same from 1930 to 1960 then N increases much more rapidly than P from 1960 to 1990
1970 to 1990 N increase, P no increase
They both increase between 1930 and 1990 but N more than P
Any ***two***

(*b*) 7·6 mg/dm^3

(*c*) As nitrate use on farmland increased, the nitrate content of the rivers also increased

13. (*d*) Causes algal blooms **or** reduction in oxygen content of water **or** death of many animals/fish **or** increase in organic matter **or** increase in numbers of bacteria
Any two

(*e*) (domestic) sewage

(*f*) eg mention of phospholipids in cell membranes **or** phosphate in ATP **or** DNA/RNA/nucleic acids

Section C

1. A. DNA

(i) **DNA structure**
1 full name
2 forms chromosomes
3 double helix/double strand/twisted **(both required)**
4 made of nucleotides **(plural important)**
5 three components named—base, deoxyribose sugar, phosphate
6 four bases named
7 complementary base pairing **(letters sufficient)**
8 sugar/phosphate backbone/phosphates linked to sugars by (strong) bonds
9 hydrogen bonds linking bases/strands
10 strands run in opposite directions

(ii) **DNA replication**
11 (hydrogen bonds) unzip + strand unwinds
12 free nucleotides attach
13 need for ATP/enzymes/DNA template
14 complementary base pairing
15 semi-conservative replication described or named/copies are identical

B. Immunity

(i) **B-lymphocytes**
1 produced in bone marrow*/mature in bone marrow
2 manufacture antibodies
3 in response to antigens **(must indicate foreign or non-self)**
4 antibody is Y shaped/protein molecule
5 specific in their action **or** described*
6 this is humoral response/antibodies are free
7 memory cells left*
8 brief description of secondary response (eg stronger/quicker)*

**Starred points can be obtained under heading of T-lymphocytes*

2. B (i) continued

T-lymphocytes
9 mature in thymus
10 this is cell-mediated response/antibodies attached
11 direct contact with infected cells (to bring about destruction)
12 T-killer and T-helper cells named.

(ii) **Macrophages**
13 non-specific in their action
14 phagocytosis named
15 description of process/engulf
16 lysosomes named
17 description of process/release enzymes

2. A. The life history of red blood cells
1 formed in bone marrow
2 iron and vitamin B_{12} needed for formation
3 pick up oxygen in lungs/deliver oxygen to tissues
4 oxygen carried by (oxy)haemoglobin
5 biconcave to increase surface area
6 for ease of oxygen transfer/diffusion
7 flexible to get through capillaries
8 reference to very small size
9 no nucleus/lack of many organelles
10 to provide more capacity for haemoglobin
11 short life span/120 days
12 destroyed in liver/spleen
13 haemoglobin broken down to produce bilirubin

B. The influence of hormones on the growth and development of boys
1 growth hormone/somatotrophin
2 released by pituitary gland
3 stimulates protein synthesis/bone growth
4 effects of over-secretion/effects of under-secretion
5 growth spurt at puberty (or age given)
6 description or drawing of growth curve
7 thyroxine affects growth
8 produced by thyroid gland **or** influenced by TSH from pituitary
9 FSH **or** LH produced by pituitary
10 FSH stimulates sperm production/LH(ISCH) stimulates production of testosterone
11 testosterone produced by testes/interstitial cells
12 testosterone stimulates development of 2nd sexual characteristics/changes in body
13 eg growth of hair/drop of voice/production of sperm/maturity/fertility/body proportions—muscle/bone changes

Higher Human Biology 2003

Section A

1. C	2. A	3. C
4. C	5. B	6. A
7. C	8. A	9. A
10. C	11. D	12. D
13. B	14. C	15. B
16. A	17. D	18. D
19. D	20. D	21. D
22. A	23. B	24. C
25. A	26. B	27. A
28. C	29. C	30. B

Section B

(Brackets indicate not essential. Solidus / indicates alternative)

1. (*a*) A cell/plasma membrane
B rough endoplasmic reticulum (rough ER)
C Golgi body/apparatus

(*b*) (i) phagocytosis/endocytosis
(ii) B transports proteins/protein synthesis
C packages/prepares/processes proteins/enzymes

(*c*) (digestive) enzymes

(*d*) (i) 1:700
(ii) The number/proportion of white cells would increase.

(*e*) biconcave shape/flattened disc provides large surface area for absorption (and/or release) of oxygen or makes it more flexible and able to pass through small capillaries

2. (*a*) (i)

Partial pressure kPa	*Percentage oxygen saturation of haemoglobin*	
	37°C	38°C
18	90	90
6	60	40
change	30	50

(ii) *Any two from*:
- Working muscles will be warmer
- which results in greater loss of oxygen from haemoglobin
- More oxygen needed for increased rate of respiration

(*b*) Oxygen is diffusing/moving (from alveoli) into the blood.

3. (*a*) Anne – $Hb^A\,Hb^A$ Philip – $Hb^A\,Hb^B$
Charles – $Hb^B\,Hb^B$

(*b*) 50%

(*c*) manufactured in bone marrow
Destroyed in liver/spleen

4. (*a*) (i) amino acid
(ii) Q - peptide/covalent
R - hydrogen

(*b*) (i) To prevent them digesting body cells in which they are manufactured.
(ii) minerals/vitamins/other enzymes/co-enzymes/hydrochloric acid

(*c*)

Factor	*Type of change*	*Effect on enzyme structure*
high temperature	**denaturation**	alters active site
α β γ or X-rays, radioactivity, UV light, mustard gas, colchicine	mutation	**alteration in amino acid sequence**

5. (*a*) 3C Pyruvic acid
6C Citric acid/tricarboxylic acid/TCA

(*b*) *line drawn between pyruvic acid and the hydrogen/CO_2 exit (X)*

(*c*) matrix of mitochondrion

(*d*)

Product	*Fate of product*
X = carbon dioxide [CO_2]	diffuses out of cell/diffuses into the blood/carried by blood to the lungs/ breathed out
Hydrogen	picked up by an acceptor/NAD carried to cytochrome system/chain combines with oxygen (to make water)

6. (*a*) Bile

(*b*)

Blood supply	*Name of blood vessel*	*Deoxygenated or oxygenated blood*
from aorta	**hepatic artery**	**oxygenated**
from gut	**hepatic portal vein**	**deoxygenated**
to vena cava	**hepatic vein**	**deoxygenated**

(*c*) liver → amino acids → urea → kidney

(*d*) Insulin, glucagon or adrenalin

7. (*a*) Volume of sample
Temperature
Starting pH
Time for experiment to take place
Size/type of tablet/mass of urease

(*b*)

Fluid sample (A, B or C)	*Ammonium carbonate concentration* (g/litre)
B	0·16
C	16·7
A	0·52

(*c*) Take more samples/repeat experiment

8. (*a*) Hypothalamus

(*b*) *Any two from*:
- shivering
- vaso-constriction
- pilo-erection/hair erection
- increased metabolic rate
- reduced sweating

(*c*) Process/mechanism brings about a return/rise of temperature to normal
Once increase has occurred the process/mechanism is switched off.

(*d*) Babies have a higher surface area to mass ratio than adults/Correct calculation of SA/MASS ratios
So babies lose relatively more heat (through their skin)

9. (*a*) 400–800

(*b*) Median volume: 1400
Percentage increase: 75%

(*c*) (i) cerebrum/cerebral hemispheres
(ii) corpus callosum
(iii) limbic system/hippocampus

10. (*a*) A dendrites
B axon

(*b*)

Feature	*Somatic*	*Autonomic*
type of control (conscious/unconscious)	**conscious/voluntary**	**unconscious/ involuntary**
example of target muscle	**any skeletal muscle eg biceps, triceps, calf muscle, thigh muscle**	uterine muscle
example of neurotransmitter	**acetylcholine**	noradrenaline

(*c*) 1 increase
2 inhibits/slows/diverts blood away
3 vasoconstriction

(*d*) Antagonistic

11. (*a*)

Label	*Type of bacteria*	*Process in nitrogen cycle*
G	**nitrogen fixing**	trap atmospheric nitrogen
J	nitrifying	**convert ammonia to nitrate**
K	**denitrifying**	convert nitrate to nitrogen gas

(*b*) increased plant/algal growth
increasing decomposition or increase in population of bacteria
decrease in oxygen leads to death of animals

12. (*a*)

	Year	*Number of cases*
Highest	1976	12200
Lowest	1999	300

(*b*) An increase in vaccination rate/introduction of vaccination programme

(*c*) 0·2

13. (*a*)

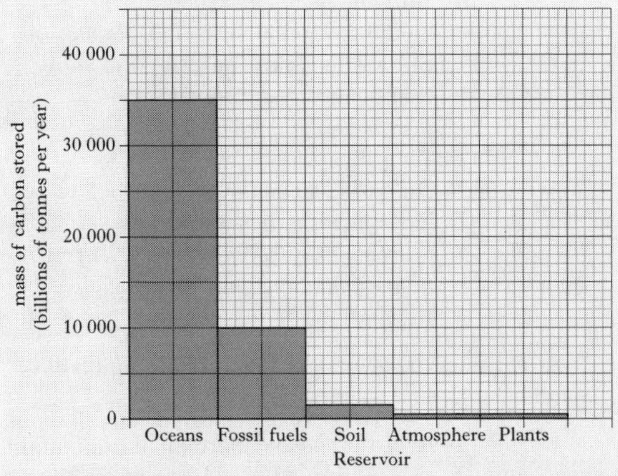

(*b*) 181 years ± 1 year (units necessary)

(*c*) *any number between 15 and 16 years acceptable*

(*d*) *Any two from*
- change in rainfall
- change in temperature or global warming
- higher wind speeds

(Your answer must refer to climate change, not to ice melt or sea level change or flooding.)

Higher Human Biology 2003 (cont.)

Section C

1. A. Discuss the influence of others on an individual's behaviour:

(i) Social facilitation

1 Performance improves in the presence of others
2 This occurs when competing with others (coactor effect)
3 This occurs when spectators are present (audience effect)
4 Suitable example, to include better performance and in the presence of others

(ii) Deindividuation

5 Individual loses personal identity/gains anonymity when in a group
6 Behaviour deteriorates
7 Individual acts as he/she would never do on own/individual takes risks
8 Suitable example, to include reference to group and poorer behaviour

(iii) Influences that change beliefs

9 Internalisation
10 Individuals change their beliefs/behaviour through persuasion
11 Suitable example described to show change in belief and source of persuasion
12 Identification
13 Individuals change their beliefs to be like someone they admire
14 Suitable example, to show change in belief/behaviour and focus of admiration

B. Exponential growth of the human population.

(i) Demographic trends

1 Rapid/explosive increase in population
2 as a consequence of birth rate exceeding death rate
3 Developing countries increasing and developed countries stable or in decline

(ii) Agriculture

4 Agricultural practice has resulted in increased food production, eg
5 fertiliser use to improve plant growth
6 pesticide use to increase yield/to kill pests
7 selective breeding described briefly
8 genetic engineering described briefly
9 improved transport/storage/preservation so less wastage
10 land reclamation/increased mechanisation improves efficiency
11 monoculture improves efficiency

1. B. (iii) Disease

12 Fewer deaths from disease because, eg.
13 improved sanitation reduces chance of infection
14 better health education results in safer behaviour
15 more hospital provision results in more treatment available
16 vaccination programmes prevent disease
17 antibiotics to treat disease
18 control of pests/disease vectors to reduce infection

2. A. Describe the influence of hormones on the testes.

1 Pituitary produces FSH
2 FSH promotes sperm production
3 Sperm are produced in the seminiferous tubules
4 Pituitary produces LH (ICSH)
5 LH stimulates testosterone production
6 in the interstitial cells
7 Testosterone stimulates sperm production
8 Testosterone influences the production of semen/prostate gland/seminal vesicles
9 Higher levels of testosterone inhibits LH/FSH
10 This is negative feedback
11 This ensures levels of testosterone are kept within normal range/constant

Relevance – for example, mention of two or more of the following will lose this mark: Oestrogen, progesterone, influence of hormones on female or mention of influence of hormones on other aspects of male non-sexual development. Any hormone unrelated to functioning of testes.

Coherence: Proper paragraphing, sequencing, sentence structure.

B. Describe the events which take place in the first half of the menstrual cycle

1 Cycle starts with menstruation
2 Breakdown of lining of uterus/endometrium
3 Pituitary gland secretes FSH
4 which stimulates growth of follicle
5 Follicle/ovary produces oestrogen
6 Oestrogen stimulates repair of endometrium/uterus lining
7 Oestrogen also stimulates production of LH
8 LH is produced by pituitary gland
9 LH brings about ovulation
10 Rising levels of oestrogen inhibit FSH production
11 This is negative feedback

Relevance – for example, mention of two or more of the following will lose this mark: Progesterone, corpus luteum, pregnancy, secondary sexual characteristics, irrelevant hormones, eg prolactin and oxytocin.

Coherence: Proper paragraphing, sequencing, sentence structure.

Higher Human Biology 2004

Section A

1. C	2. A	3. D
4. B	5. D	6. D
7. D	8. A	9. B
10. B	11. D	12. C
13. A	14. C	15. A
16. C	17. A	18. A
19. B	20. C	21. C
22. D	23. B	24. D
25. C	26. A	27. A
28. B	29. B	30. C

Section B

1. (*a*) $CO_2 + H_2O$
ADP + P_i
amino acids or (poly)peptides

(*b*) (i) Glycolysis in cytoplasm
or
Krebs cycle in mitochondrion/matrix of mitochondrion
or
Cytochrome/electron transport chain in mitochondrion/cristae of mitochondrion
(ii) Ribosome

(*c*) No oxygen, Lactic acid instead of $CO_2 + H_2O$, No pathway X

(*d*) Lipid, fat, fatty acid, glycerol, protein, amino acid.

2. (*a*)

	H	h
H	HH	Hh
h	Hh	hh

(*b*) 1. 25% 2. 50% 3. 25%

3 (*a*)

Type of cell	*Secretion of antibodies*	*Type of response*
B-lymphocyte	yes	Humoral
T-Lymphocyte Killer T-cell	no	cell-mediated response
Macrophage or Monocyte or Neutrophil	no	non-specific response

(*b*) *Arrow should be pointing downwards, towards the lymph node.*
The valve would close if fluid was flowing in the other direction *or similar*

(*c*) It has less proteins in it **or** it has more lipids/fats in it

(*d*) It rejoins the main circulatory system/blood

(*e*) Removal of excess tissue fluid **or** transport of lipids/fats (from digestive system)

4. (*a*) 136/82

(*b*) That caffeine causes an increase in blood pressure

(*c*) Students given a cup of decaffeinated coffee (of same strength and volume) **or** water

(*d*) *Any two from:*
- Volume/strength of coffee/time taken to drink coffee
- Mass of caffeine in drink
- Activity before and after taking readings **or**
- Activity during the hour between readings
- Ingestion of other food

(*e*) Systolic – (maximum pressure) when heart is contracting **or** pressure to stop blood flow in artery.
Diastolic – (minimum pressure) when heart is relaxing **or** pressure to allow free flow of blood in artery.

(*f*) (i) 3
(ii) 15

5. (*a*) A. Oviduct/fallopian tube
B. Ovary

(*b*) Cleavage

(*c*) Oestrogen/progesterone produced by ovary **or** follicle/corpus luteum

(*d*) Placenta

(*e*) Monozygotic twins develop from one egg fertilised by one sperm. The zygote/ball of cells then divides into two separate individuals.
Dizygotic twins develop from two eggs fertilised by two sperm.

6. (*a*) 33%

(*b*) 145

(*c*) Because the concentration of insecticide increases as it is passed along the food chain

(*d*) FSH

(*e*) Prostate, Cowper's gland **or** seminal vesicle.
Nourishment/activation/source of sugar (fructose) for sperm
Stimulates contraction of the uterus/provides a medium for sperm to swim in.

(*f*) Artificial insemination – Sperm injected into womb by syringe/mechanically/by method other than sexual intercourse
In vitro fertilisation – fertilisation takes place outside the uterus/in glass dish/test tube

7. (*a*) Polygenic

(*b*) ABO

(*c*) Cell membrane/plasma membrane

(*d*) If father is homozygous (DD), child will be Rh +ve
If father heterozygous (Dd), child has 50% chance of being Rh +ve

(*e*) (Exchange) blood transfusion of the baby can be carried out **or** mother can be injected with anti-D antibody/immunoglobin

Higher Human Biology 2004 (cont.)

8. (*a*) 1236

(*b*) 6435 ± 1

(*c*) The immune system recognises the transplant tissue as foreign (Without suppression,) it would attack the transplant and destroy it

(*d*) Renal artery and vein

9. (*a*) Cerebrum/cerebral hemispheres/cortex

(*b*) To control muscle movement/to send impulses to effectors/muscles

(*c*) The hands have much more fine motor control than the feet

(*d*) Diverging

(*e*) Non-verbal communication

10. (*a*) Axon

(*b*) (i) Neurotransmitter

(ii) An (electrical) impulse/ionic change

(iii) It is removed by enzyme action

(*c*) Q – myosin
R – actin

(*d*) They slide together/they overlap more

11. (*a*) The population which can be sustained by an ecosystem/environment

(*b*) (i) 10:1

(ii) *Any two from:*

- Over-fishing/better boats/better hunting/detection equipment
- An increase in pollution/disease
- Climatic change affecting temperature of the water

12. (*a*) (i) Birth rate – 44
Death rate – 37

(ii) Introduction of contraception/improved economic conditions/emancipation of women/improved education/reduced male fertility

(iii) Improved sanitation/health service/diet **or** introduction of vaccination programme

(*b*) Increasing – Because death rate is always lower than birth rate

(*c*) Country A – *Any one from:*

- because the rate of population increase is increasing
- because the birth rate is relatively high
- because the death rate is relatively high

13. (*a*)

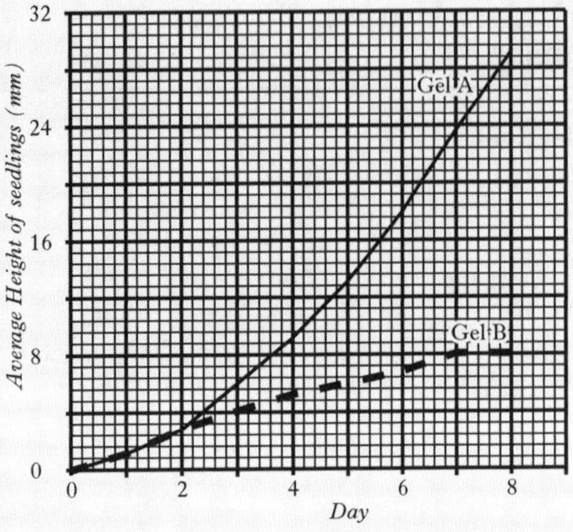

(*b*) The absence of nitrate inhibits growth/nitrates stimulate growth

(*c*) The control is tube A.
It contains the complete set of minerals to show how plants grow under normal conditions

(*d*) More than one seed was grown in each condition

(*e*) To make sure there were no nitrates in the water

(*f*) eg DNA/RNA *(**or** any other compound containing P)*

Section C

1. A.

(i) *Any of the following:*

1. Capacity – around 7 pieces of information (± 1)
2. Held for a very short period of time/seconds only
3. This is memory span
4. Chunking + example of chunking
5. Serial position effect **or** description of SPE
6. Encoding + description (eg acoustic, semantic, visual)
7. Located in limbic system/hippocampus (can be gained in either (i) **or** (ii))

(ii) *Any of the following:*

8. Rehearsal
9. Repetition of items to be memorised
10. Organisation
11. Putting items into groups **or** categories
12. Elaboration
13. Putting meaning to items/adding additional information to items
14. Mnemonics (memory aids/systems) + description **or** example
15. Memory maps/mind mapping/visual imagery + description **or** example

1. B.

(i) *Any of the following, each with short description*
1. Irrigation
2. Terracing **or** increase in size of fields
3. Crop rotation **or** monoculture
4. Deforestation
5. Drainage/flood control
6. Reclamation (sea/desert/marginal land)
7. Mechanisation

(ii) *Any of the following:*
8. Fertilisers
9. Function – promotes growth of plants
10. Examples – nitrates/phosphates etc
11. Pesticides
12. Fungicides kill fungi/moulds
13. Herbicides/weedkillers kill weeds
14. Insecticides kill insects
15. Antibiotics to protect animals against disease
16. Hormones + simple description in plant **or** animal husbandry eg growth promoters.

2. A. *Any of the following:*
1. Meiosis as nuclear/cell/reduction division to produce sex cells/sperm and eggs *(ie definition of meiosis which includes knowledge of what a gamete is)*
2. Pairing of homologous chromosomes
3. Crossing over occurs
4. Chromatids break and rejoin/genetic exchange *(labelled diagram accepted)*
5. These points are called chiasmata
6. Chromosomes line up along equator
7. Assortment is independent **or** explanation *(labelled diagram accepted)*
8. Describe numerically. e.g. 23 pairs gives over 8 million different combinations
9. Non-disjunction
10. This is a form of mutation
11. Description of non-disjunction
12. Example of non-disjunction - eg Down's syndrome

B. *Any of the following:*
1. mRNA takes code from DNA/nucleus to...
2. site of activity - ribosomes **or** rough ER
3. mRNA contains codons
4. tRNA contains an anticodon
5. a codon/anticodon consists of three bases
6. tRNA carries a specific amino acid *(labelled diagram accepted)*
7. codons match/link up/pair with anticodons
8. bases match, ie. U to A and G to C *(both pairings required – diagram accepted)*
9. results in amino acids lining up
10. peptide bonds join amino acids to make peptide/polypeptide/protein
11. the sequence of bases/codons = sequence of amino acids
12. there is a requirement for ATP and enzymes

Official SQA answers to 1-84372-243-7
2000–2004

2000 HIGHER

FOR OFFICIAL USE

Total

X009/301

NATIONAL QUALIFICATIONS 2000

MONDAY, 29 MAY
9.00 AM – 11.30 AM

HUMAN BIOLOGY HIGHER

Fill in these boxes and read what is printed below.

Full name of centre

Town

Forename(s)

Surname

Date of birth
Day Month Year

Scottish candidate number

Number of seat

SECTION A—Questions 1–30

Instructions for completion of Section A are given on page two.

SECTIONS B AND C

1 (a) All questions should be attempted.

(b) Question 11 is on pages 20, 21 and 22 and Question 12 is on pages 23, 24 and 25—pages 22 and 23 are fold-out pages.

(c) It should be noted that in **Section C** questions 1 and 2 each contain a choice.

2 The questions may be answered in any order but all answers are to be written in the spaces provided in this answer book, and must be written clearly and legibly in ink.

3 Additional space for answers and rough work will be found at the end of the book. If further space is required, supplementary sheets may be obtained from the invigilator and should be inserted inside the **front** cover of this book.

4 The numbers of questions must be clearly inserted with any answers written in the additional space.

5 Rough work, if any should be necessary, should be written in this book and then scored through when the fair copy has been written.

6 Before leaving the examination room you must give this book to the invigilator. If you do not, you may lose all the marks for this paper.

SCOTTISH QUALIFICATIONS AUTHORITY

SECTION A

Read carefully

1 Check that the answer sheet provided is for Human Biology Higher (Section A).

2 Fill in the details required on the answer sheet.

3 In this section a question is answered by indicating the choice A, B, C or D by a stroke made in **ink** in the appropriate place in the answer sheet—see the sample question below.

4 For each question there is only **one** correct answer.

5 Rough working, if required, should be done only on this question paper—or on the rough working sheet provided—**not** on the answer sheet.

6 At the end of the examination the answer sheet for Section A **must** be placed **inside** this answer book.

Sample Question

The digestive enzyme pepsin is most active in the

A mouth

B stomach

C duodenum

D pancreas.

The correct answer is **B**—stomach. A **heavy** vertical line should be drawn joining the two dots in the appropriate box in the column headed **B** as shown in the example on the answer sheet.

If, after you have recorded your answer, you decide that you have made an error and wish to make a change, you should cancel the original answer and put a vertical stroke in the box you now consider to be correct. Thus, if you want to change an answer D to an answer B, your answer sheet would look like this:

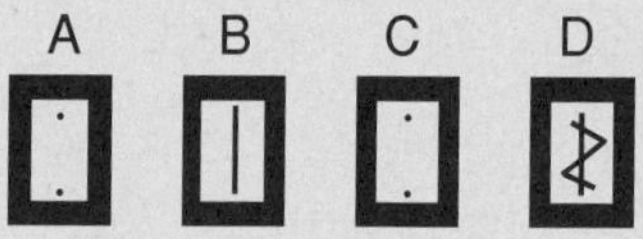

If you want to change back to an answer which has already been scored out, you should enter a tick (✓) to the **right** of the box of your choice, thus:

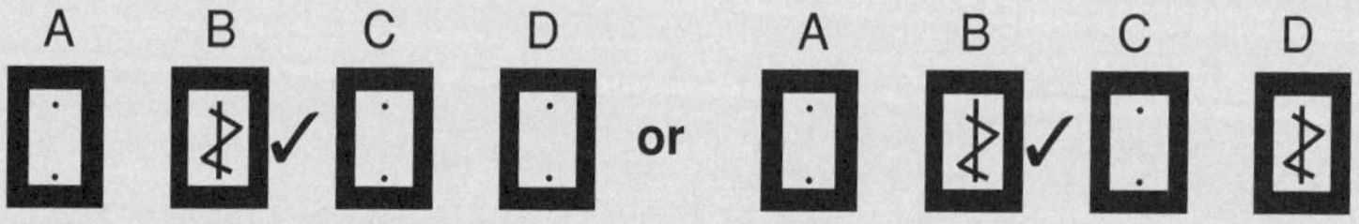

SECTION A

All questions in this section should be attempted.

Answers should be given on the separate answer sheet provided.

1. What is the name of the cell structure shown in the diagram below?

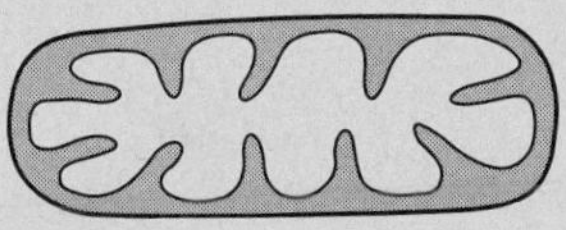

A Golgi body

B Mitochondrion

C Lysosome

D Ribosome

2. Which of the following tissues is rich in both actin and myosin?

A Muscle tissue

B Liver tissue

C Nerve tissue

D Adipose tissue

3. Endocytosis is best described as

A the uptake of a substance by a cell by active transport

B the export of a substance through a cell membrane

C the uptake of a substance in a vesicle formed by the cell membrane

D the diffusion of a substance along a concentration gradient.

4. The specificity of an enzyme is determined by the

A presence of an inhibitor

B substrate concentration

C state of equilibrium of the reaction

D molecular structures of substrate and enzyme.

5. Which of the following cells secrete antibodies?

A Bacteria

B Macrophages

C T lymphocytes

D B lymphocytes

6. The table refers to the mass of DNA in certain human body cells.

Cell type	*Mass of DNA in cell* [$\times 10^{-12}$ g]
liver	6·6
lung	6·6
P	3·3
Q	0·0

Which of the following is the most likely identification of cell types P and Q?

	P	*Q*
A	kidney tubule cell	ovum
B	ovum	mature red blood cell
C	mature red blood cell	sperm
D	nerve cell	mature red blood cell

7. The mRNA codon for the amino acid threonine is ACU. What is the corresponding anti-codon?

A ACT

B UCT

C UGA

D TGA

8. Haploid gametes are produced during meiosis as a result of

A the separation of homologous chromosomes

B the independent assortment of chromosomes

C the separation of chromosomes into chromatids

D the crossing over of chromatids.

9. The transmission of a gene for deafness is shown in the family tree below.

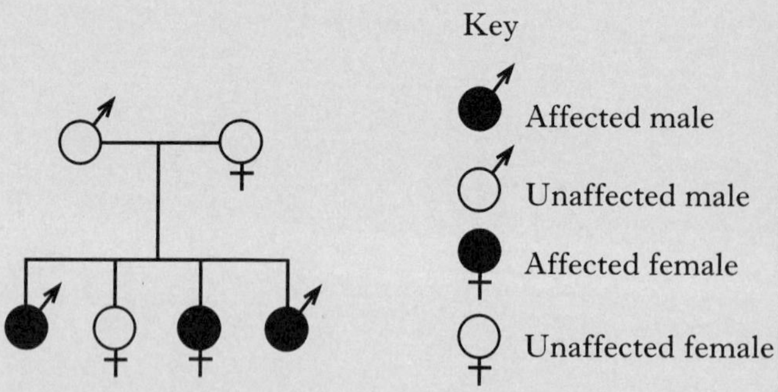

This condition is controlled by an allele which is

A dominant and not sex-linked

B recessive and not sex-linked

C dominant and sex-linked

D recessive and sex-linked.

10. Red-green colour blindness is a sex-linked recessive trait. A woman whose father is colour blind marries a man with normal vision. If they have a daughter, what are the chances she will be colour blind?

A 0%

B 25%

C 33%

D 50%

11. Identical twins can result from

A two haploid eggs fertilised by two identical sperm

B a haploid egg fertilised by two identical sperm

C a diploid egg fertilised by a single sperm

D a haploid egg fertilised by a single sperm.

12. In a normal individual, which of the following gametes cannot be formed?

A A sperm with an X chromosome

B A sperm with a Y chromosome

C An egg with an X chromosome

D An egg with a Y chromosome

13. The cell shown is magnified three hundred times. What is the actual size of the cell?

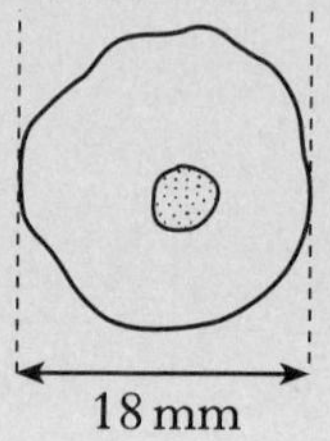

A 6 µm

B 60 µm

C 54 µm

D 540 µm

14. The diagram below records the beat of a human heart.

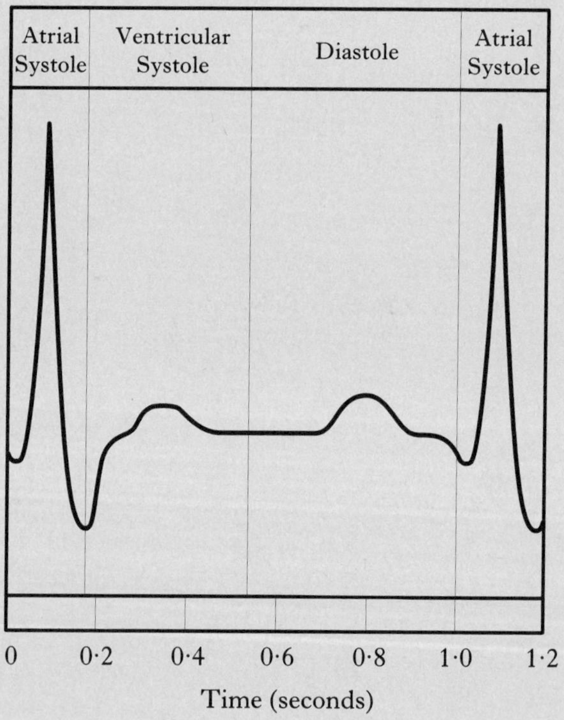

What is the heart rate in beats per minute (bpm)?

A 50 bpm

B 60 bpm

C 70 bpm

D 120 bpm

15. Which of the events below produces the normal sounds of heart beat heard through a stethoscope?

A Contraction of the atria

B Contraction of the ventricles

C Closing of the heart valves

D Opening of the heart valves

16. A woman with blood group *AB* has a child to a man with blood group *O*. What are the possible phenotypes of the child?

A *AB* or *O*

B *A* or *B*

C *AB* only

D *AB*, *A* or *B*

17. The diagram below shows certain organs found in the abdomen.

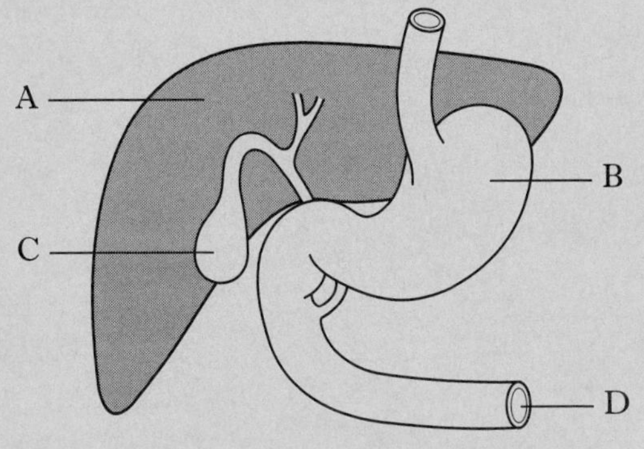

Where is bile active?

18. Which of the following correctly describes the flow of blood in the blood vessels associated with the liver?

	Hepatic vein	*Hepatic artery*	*Hepatic portal vein*
A	from liver	to liver	from gut
B	from liver	to gut	from gut
C	to liver	from liver	to gut
D	to liver	to gut	from liver

19. Which of the following pairs of compounds is produced by the pancreas?

A Glycogen and insulin

B Insulin and ADH

C Insulin and glucagon

D Glycogen and glucagon

20. The table below records the concentration of urea in plasma and urine.

	Plasma	*Urine*
Urea [g/100 cm^3]	0·2	1·26

By how many times has the urea been concentrated by the activity of the kidney?

A 0·252 times

B 1·28 times

C 6·3 times

D 63·0 times

21. Which of the following hormones is released by the pituitary gland?

A Insulin

B Adrenaline

C Oestrogen

D Antidiuretic hormone

22. Infants are more likely to suffer from hypothermia because they have

A a low surface area to volume ratio

B a high surface area to volume ratio

C a low metabolic rate

D a high metabolic rate.

[Turn over

23. The diagram below shows a section through a brain.

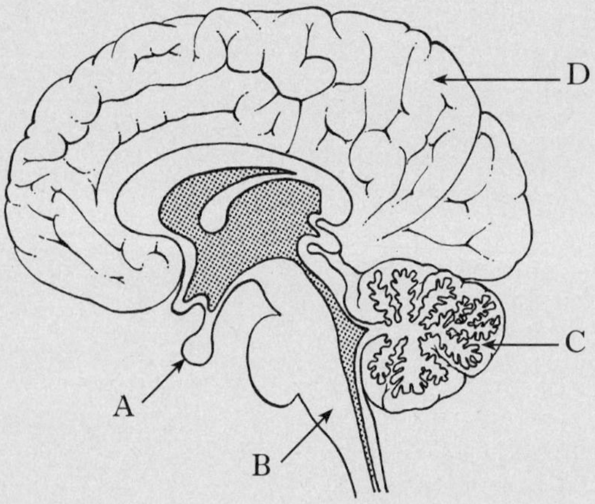

Which letter indicates the part of the brain which controls reflex actions?

24. The transformation of information into a form that the memory can accept is called

A encoding

B shaping

C rehearsal

D storage.

25. The rewarding of patterns of behaviour which approximate to desired behaviour is called

A generalisation

B discrimination

C shaping

D imitation.

26. The world production of wheat and soybean between 1981 and 1991 is shown on the graph below.

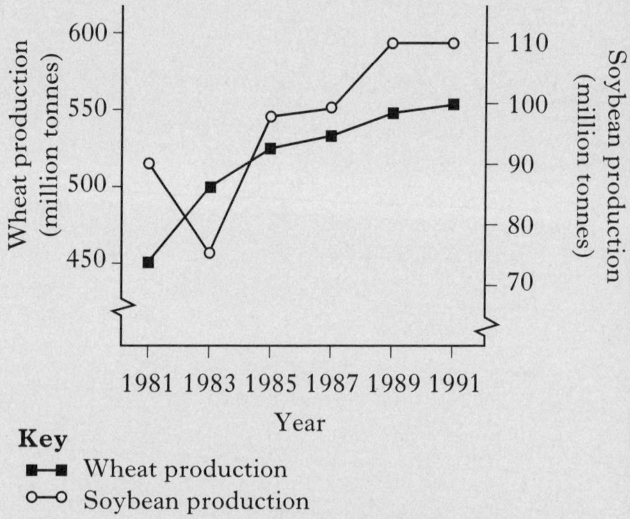

Which of the following statements relating to the data is correct?

A Wheat production is 5 times greater than soybean production in 1991.

B Soybean production is 10% greater than wheat production in 1991.

C Wheat production is 10% less than soybean production in 1981.

D Soybean production is greater than wheat production in 1981 and 1991.

27. A country has a population of 50 million. What is the likely increase in population over a two year period, given a growth rate of 1% per annum?

A 1 000 000

B 1 005 000

C 1 050 000

D 2 000 000

28. Which two gases contribute most to global warming?

A Oxygen and carbon dioxide

B Ozone and carbon dioxide

C Methane and ozone

D Methane and carbon dioxide

29. Which of the following processes results in the addition of carbon dioxide to the atmosphere?

A Decomposition

B Photosynthesis

C Ozone depletion

D Nitrogen fixation

30. Rivers polluted by raw sewage have low oxygen concentrations as a direct result of

A large numbers of bacteria

B algal blooms

C fertiliser run-off

D low nutrient levels.

Candidates are reminded that the answer sheet MUST be returned INSIDE this answer book.

[Turn over for Section B on *Page eight*

DO NOT WRITE IN THIS MARGIN

Marks

SECTION B

All questions in this section should be attempted.

1. The formation of a mRNA strand on a section of a chromosome is shown in the diagram below.

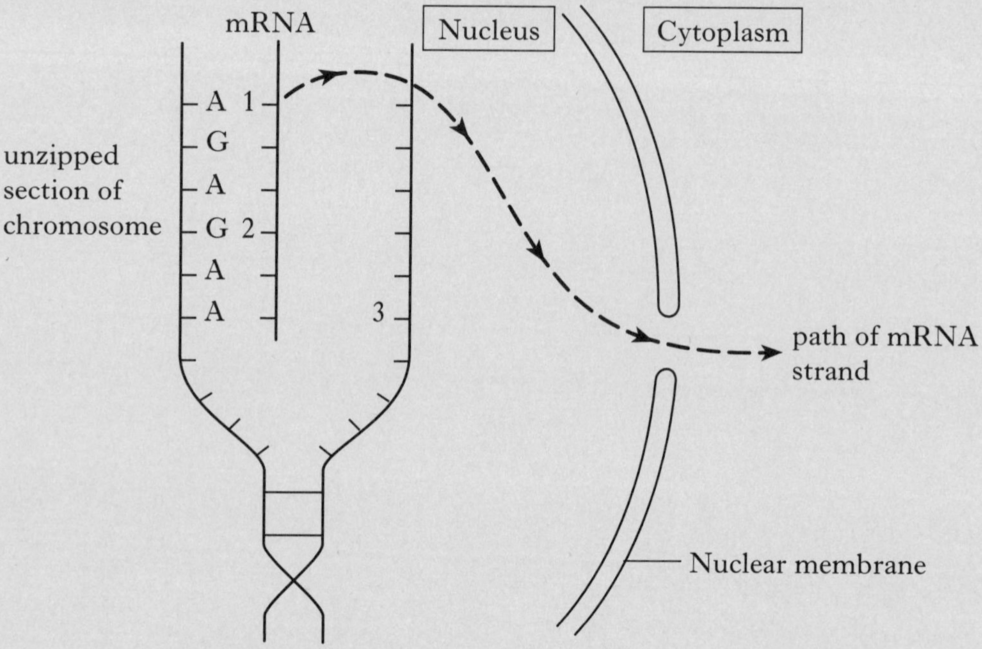

(*a*) Give the names of bases 1, 2 and 3.

1 ______________ 2 ______________ 3 ______________ **1**

(*b*) The mRNA strand is constructed from free nucleotides.

Name the **two** molecules which combine with a base to form a mRNA nucleotide.

1 ______________ 2 ______________ **1**

(*c*) Once completed, the mRNA strand moves into the cytoplasm.

What is its destination?

______________ **1**

(*d*) Explain why the formation of mRNA strands is essential to cell metabolism.

____

____

______________ **2**

DO NOT WRITE IN THIS MARGIN

Marks

2. (a) The table below contains three statements about two stages of cellular respiration.

Complete the table to indicate whether the statements are True (**T**) or False (**F**) for each stage.

	Stages of Respiration	
Statement	*Glycolysis*	*Cytochrome System*
Occurs in the mitochondrion		**T**
Releases carbon dioxide	**F**	
Uses oxygen		

2

(b) The diagram below summarises anaerobic respiration in a muscle cell.

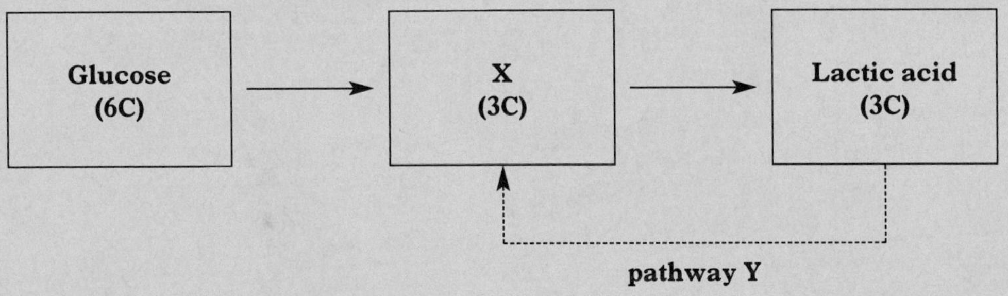

(i) Name substance X.

____________________ 1

(ii) Which substance would need to be present for pathway Y to occur?

____________________ 1

(iii) Why is anaerobic respiration considered to be a less efficient process than aerobic respiration?

____________________ 1

(iv) Glucose is not stored in muscle cells.

Name the carbohydrate which is stored in muscle cells.

____________________ 1

[Turn over

DO NOT WRITE IN THIS MARGIN

Marks

3. The diagram below represents the structure of the virus which causes influenza.

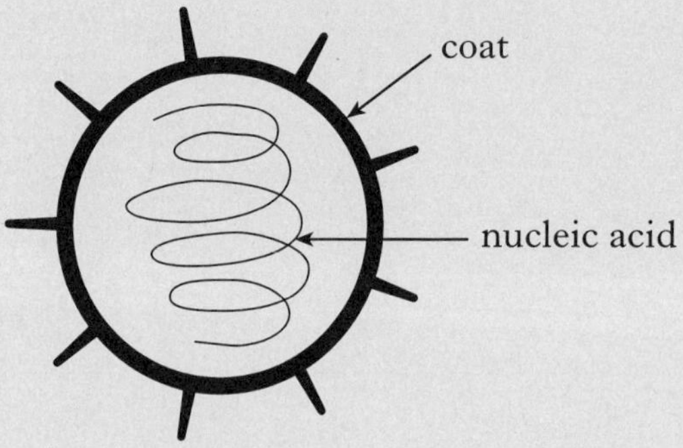

(*a*) (i) Name the substance which makes up the coat of this virus.

____________________ **1**

(ii) Describe how a virus from an influenza vaccine might differ in structure from the virus shown in the diagram.

____

____________________ **1**

(*b*) The following diagrams show a cell being attacked by the influenza virus.

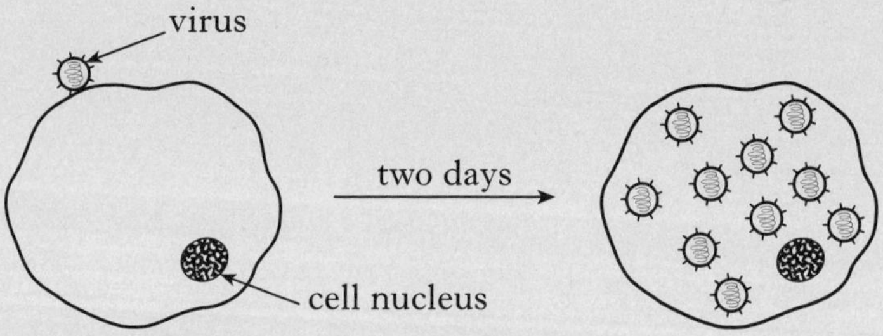

(i) Describe how the virus reproduces over the two day period.

____

____

____________________ **2**

(ii) State how viruses are released from the infected cell.

____________________ **1**

(*c*) There are three main types of influenza virus.

Why are the antibodies produced against one type not effective against the others?

____

____________________ **1**

DO NOT WRITE IN THIS MARGIN

Marks

3. (continued)

(*d*) The deaths in Scotland caused by influenza, pneumonia and bronchitis, between 1976 and 1982 are shown on the graph below.

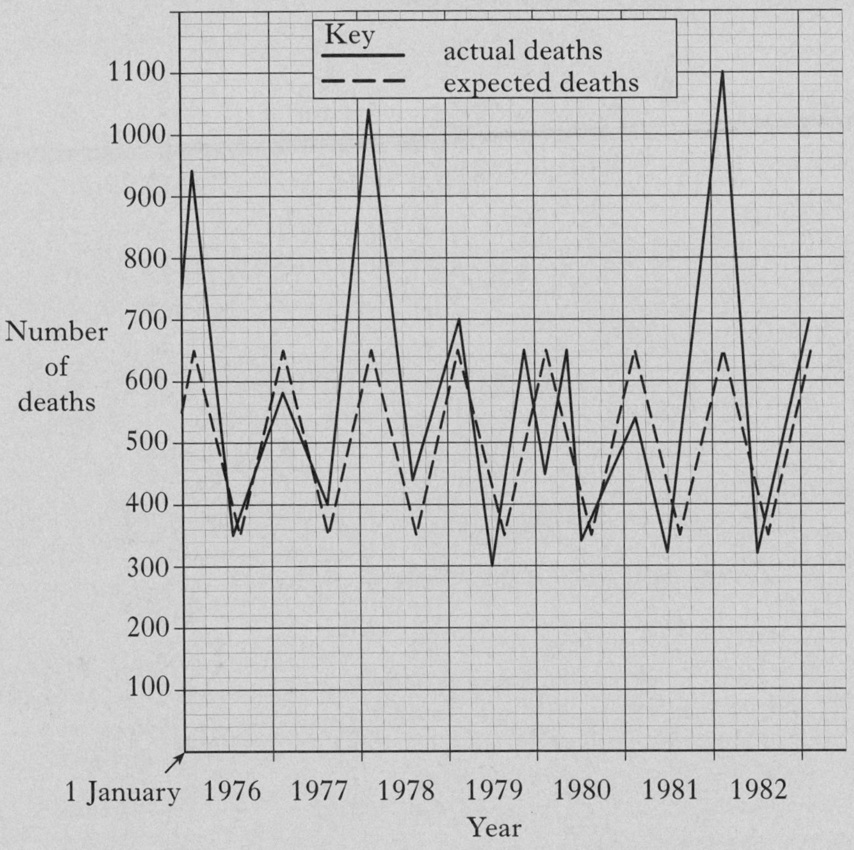

(i) What is the expected range of deaths in any year?

________________ to ________________ **1**

(ii) Suggest in which years influenza epidemics occurred in Scotland.

________________________________ **1**

(iii) Chilling reduces the action of the cilia in the windpipe.

Explain why this makes people more vulnerable to influenza in the winter.

____

________________________________ **1**

(iv) Suggest why the actual deaths in 1980 did not have the same pattern as seen in other years.

____

________________________________ **1**

DO NOT WRITE IN THIS MARGIN

Marks

4. Achondroplasia is an autosomal disorder of bone growth that is caused by a gene mutation. A person with this condition has relatively short arms and legs.

The incidence of achondroplasia in a family is shown in the diagram below.

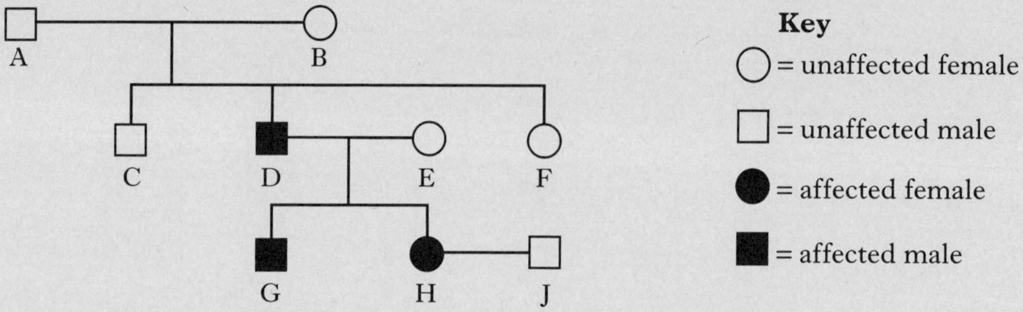

(*a*) (i) In this family, the mutation occurred in an ovum.

In which individual did the mutation take place?

________________ **1**

(ii) Describe a change which takes place in a gene when it mutates.

____

____ **1**

(*b*) The allele which causes achondroplasia is dominant.

What is the percentage chance that a child of H and J would inherit achondroplasia?

Space for calculation

____________ % **1**

(*c*) People who are affected by achondroplasia do not produce enough growth hormone. State the site of growth hormone production.

______________________________ **1**

(*d*) The list below contains a number of inherited disorders.

Phenylketonuria	**Down's syndrome**	**Huntington's chorea**
Haemophilia	**Cystic fibrosis**	

Select a disorder from this list which is:

(i) the result of non-disjunction; ____________________________

(ii) a sex-linked condition. ____________________________ **2**

DO NOT WRITE IN THIS MARGIN

Marks

5. The diagram shows a section through the reproductive organs of a man.

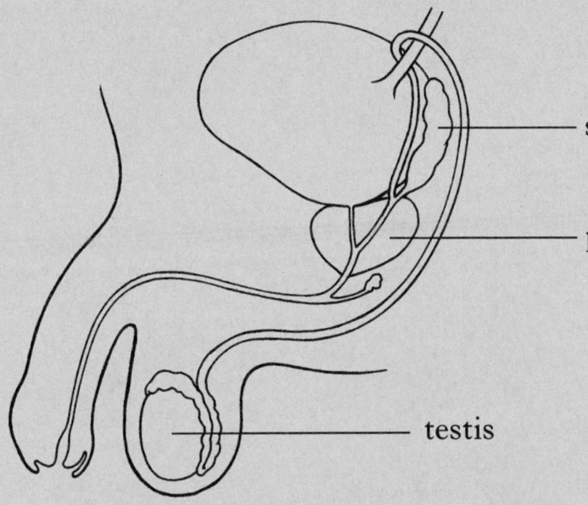

(*a*) (i) State the site of sperm production within the testis.

_______________ **1**

(ii) State **one** function of the secretions from the seminal vesicles and prostate gland.

____

_______________ **1**

(*b*) (i) During a male sterilisation operation (vasectomy), a tube is cut.

Draw a letter X on the diagram to indicate the likely position of the cut. **1**

(ii) Why is the transport of testosterone from the testes unaffected by this operation?

_______________ **1**

[Turn over

DO NOT WRITE IN THIS MARGIN

Marks

6. The concentration of progesterone in the blood of a woman over a 40 week period of time during which she became pregnant is shown on the graph below.

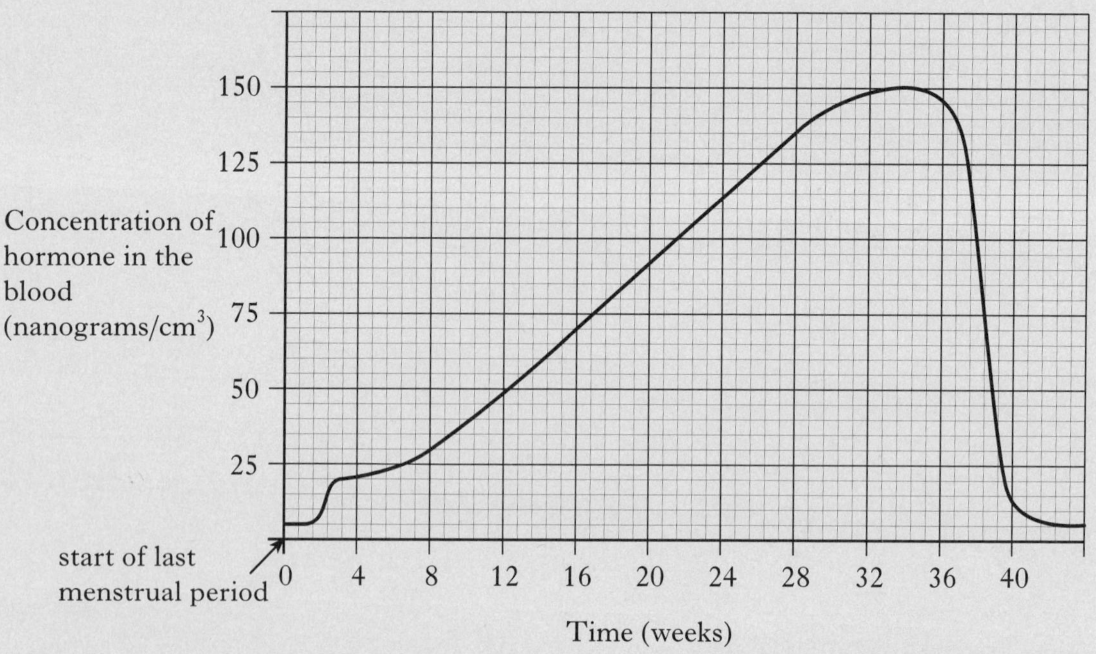

(*a*) Draw a vertical line on the graph to indicate the time of ovulation. **1**

(*b*) Describe the changes in progesterone concentration during the first 32 weeks.

____

____

______________________________ **2**

(*c*) Where is progesterone produced

(i) during the first weeks of pregnancy? ______________ **1**

(ii) during the later stages of pregnancy? ______________ **1**

(*d*) The decrease in the level of progesterone shown on the graph stimulates the production of prolactin.

What is the effect of prolactin?

____

______________________________ **1**

DO NOT WRITE IN THIS MARGIN

Marks

7. The diagram below shows the relationship between blood, tissue fluid and lymph.

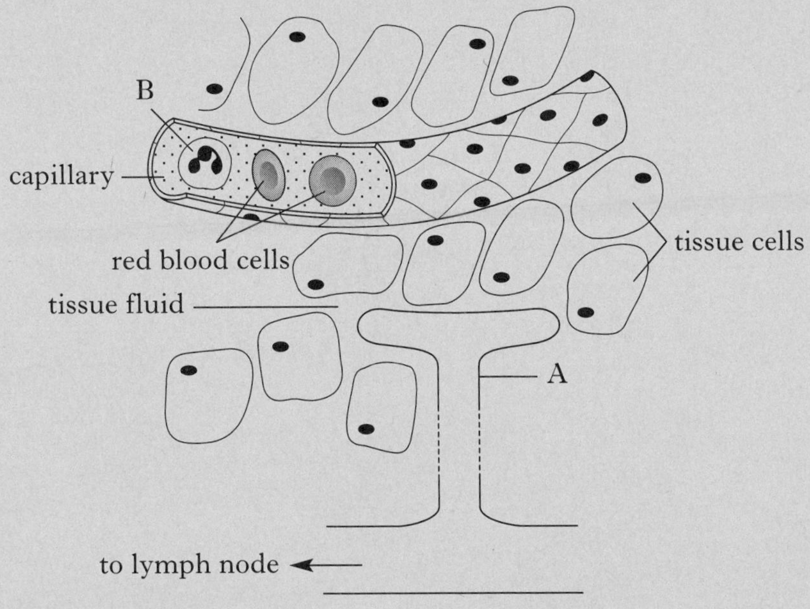

(*a*) What is the function of vessel A? **1**

(*b*) Describe how the flow of fluid in vessel A is maintained. **2**

(*c*) Explain why the lymph node may become swollen if the tissues are invaded by bacteria. **1**

(*d*) Cell B contains many lysosomes. What is the function of cell B? **1**

(*e*) How does oxygen from the red blood cells reach the tissue cells? **1**

(*f*) Explain why blood pressure falls as the blood flows through the capillaries. **1**

DO NOT WRITE IN THIS MARGIN

Marks

8. The relationship between skin temperature and sweating rates of an individual in different environmental conditions is shown on the graph below.

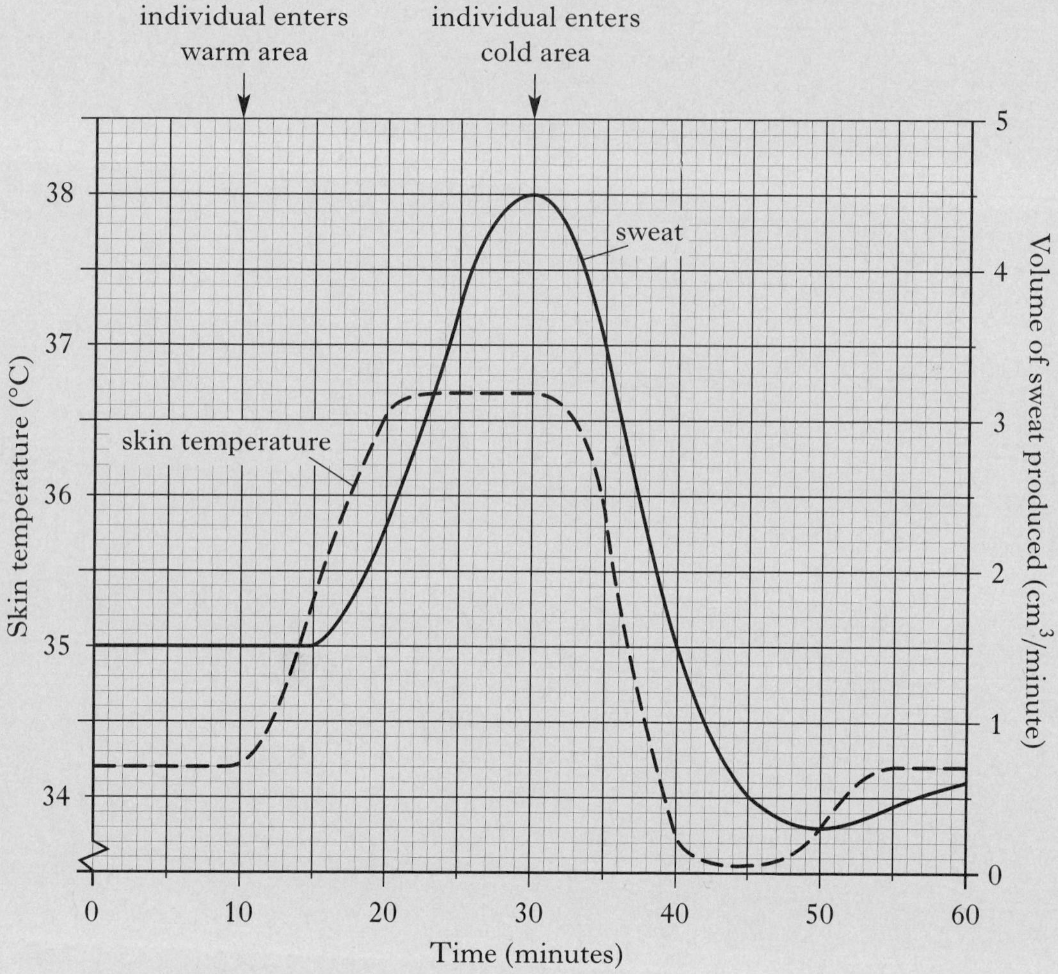

(*a*) (i) Express, as a simple ratio, the volume of sweat produced when entering the warm area to the volume of sweat produced when leaving the warm area.

Ratio ______________ : ______________
entering warm area   leaving warm area **1**

(ii) What is the relationship between skin temperature and sweat production?

____

______________________________ **1**

(iii) If the individual had remained within the warm area, predict the skin temperature at 45 minutes.

______________ **1**

DO NOT WRITE IN THIS MARGIN

Marks

8. (*a*) (continued)

(iv) What evidence from the graph suggests that sweat is produced as a result of changes in the skin temperature? **1**

(*b*) Apart from sweating, state **one** other involuntary response to an increase in body temperature. **1**

(*c*) Where in the brain is the temperature monitoring centre located? **1**

[Turn over

DO NOT WRITE IN THIS MARGIN

Marks

9. The diagram shows how the nervous system is organised.

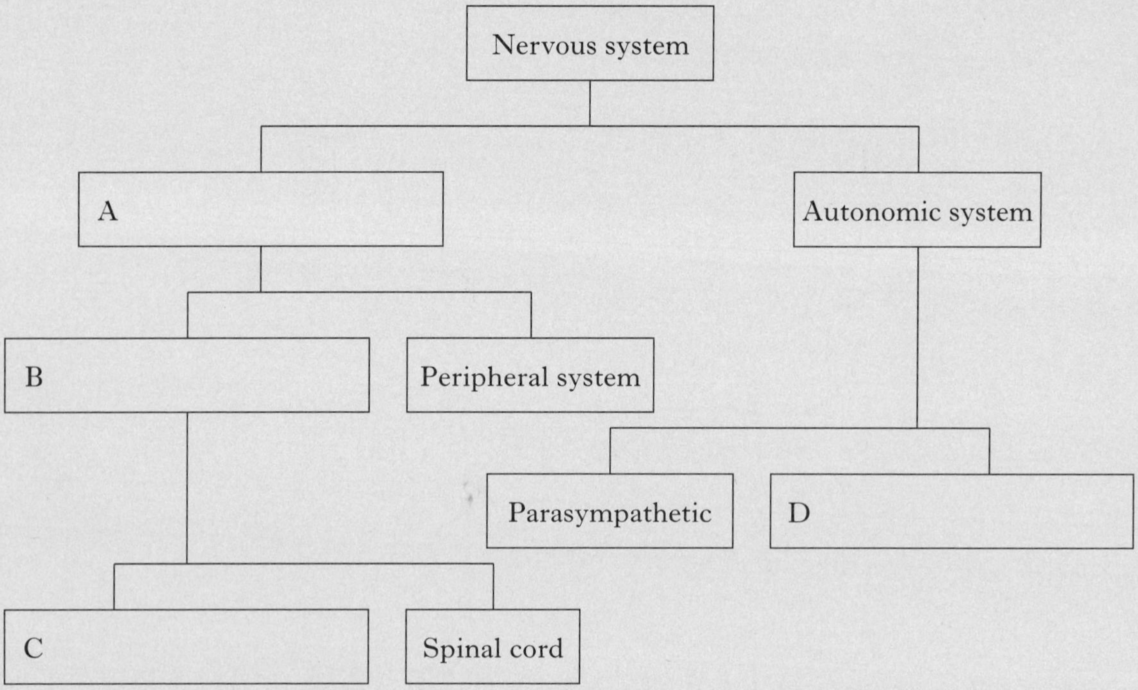

(*a*) Complete the diagram by entering the names of parts A to D. **2**

(*b*) The parts of the autonomic nervous system are described as antagonistic.

(i) What is meant by the term *antagonistic*?

____

____ **1**

(ii) Explain how this antagonistic action controls the activity of the digestive system.

____

____

____

____ **2**

DO NOT WRITE IN THIS MARGIN

Marks

10. Rods and cones are light receptor cells found in the eye.

The diagram below represents the arrangement of these cells.

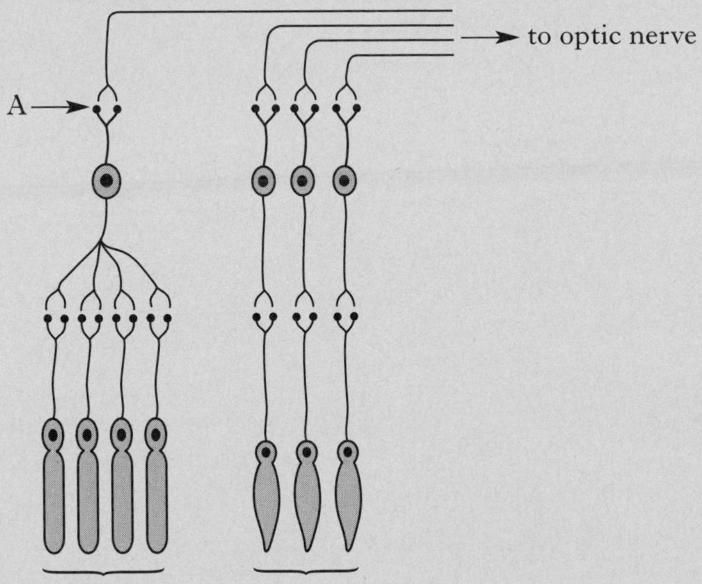

(*a*) In which region of the eye are rods and cones located?

______________________________ **1**

(*b*) (i) What type of neural pathway links the rod cells to the optic nerve?

______________________________ **1**

(ii) Name gap A.

______________________________ **1**

(iii) Explain how the arrangement of rod cells increases the chances of a nerve impulse crossing gap A.

____

____

______________________________ **2**

(*c*) The optic nerve carries nerve impulses to the visual area of the brain.

In which region of the brain is the visual area found?

______________________________ **1**

[Turn over

DO NOT WRITE IN THIS MARGIN

Marks

11. An experiment was carried out to investigate the transfer of information between short-term and long-term memory. Two groups of ten students were shown word lists for 30 seconds. The students then watched a video for 20 minutes, before being asked individually to recall as many words as possible from the list. The two lists and the results of the investigation are shown below.

Experiment one

List 1 (shown to group 1)	*List 2 (shown to group 2)*
spoon fork knife plate pencil ruler rubber pen bus train plane car potato carrot beetroot turnip	spoon pen car turnip ruler plane knife beetroot plate train pencil potato bus fork rubber carrot
Average number of words recalled 13·6	Average number of words recalled 7·8

(*a*) What feature of this investigation ensures that the results are reliable?

____

____ **1**

(*b*) Why did the students recall more words from list 1 than list 2?

____

____ **1**

[Question 11 continues on *Page twenty-one* and fold-out *Page twenty-two*

11. (continued)

(*c*) In another experiment, list 2 was read out, without pauses, to a third group of 10 students who were then asked to recall the words.

The results are shown in the table below.

Experiment two

Position of word in the series	*Percentage of correct responses*
1	100
2	90
3	100
4	90
5	80
6	80
7	50
8	40
9	60
10	70
11	80
12	100
13	90
14	100
15	100
16	100

[Question 11 continues on *Page twenty-two*

Marks

11. (*c*) (continued)

(i) Construct a line graph to illustrate the data in the table.

(Additional graph paper, if required, will be found on page 27)

2

(ii) Explain why words at the **beginning** of the list were remembered better than those in the middle of the list.

____

____ **1**

(iii) Explain why words at the **end** of the list were remembered better than those in the middle of the list.

____

____ **1**

(iv) What term is used to describe the results obtained in experiment two?

______________________ **1**

(*d*) What aspects of experimental design limit rehearsal in experiment one and experiment two?

Experiment one ______________________________________

Experiment two ______________________________________ **2**

[Question 12 begins on *Page twenty-three*

12. This question relates to causes of death in developed and developing countries in 1997.

Figure 1 Main causes of death in 1997.

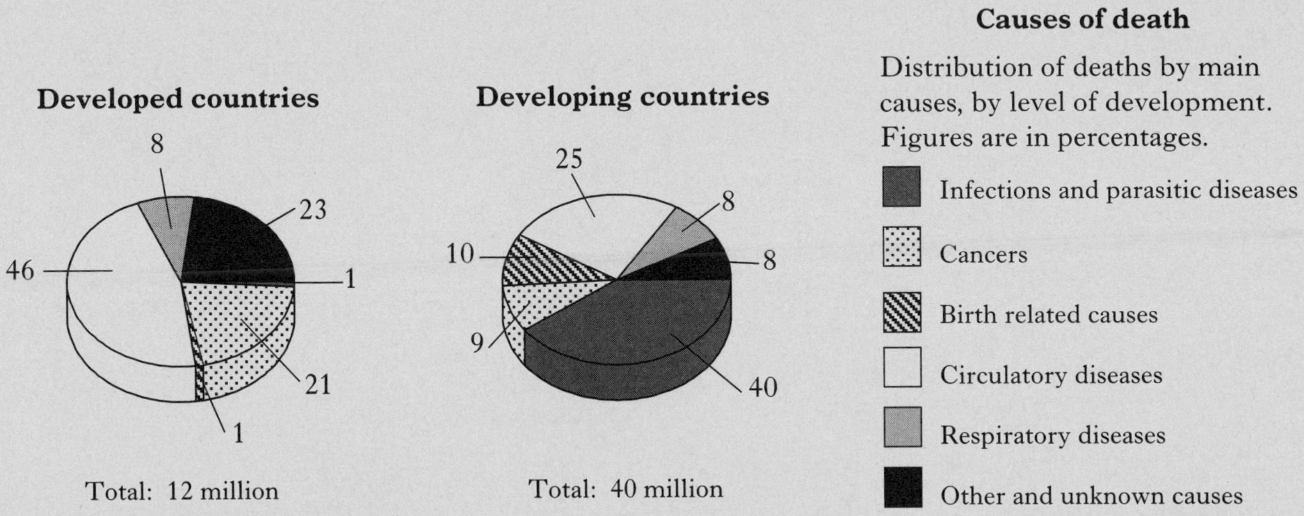

Figure 2 Main causes of death of children aged under three years in developing countries.

Cause of death	*Specific cause*	*Percentage of total deaths*
Respiratory diseases	Infections such as pneumonia or bronchitis	20·0
Infections and parasitic diseases	Diarrhoea Measles Malaria Whooping cough	19·1 10·4 6·7 3·8
Birth related causes	Prematurity Birth asphyxia Congenital abnormalities Birth trauma Others	9·5 8·6 4·8 3·8 7·6
Other	Malnutrition Other, including tuberculosis	2·8 2·9

Figure 3 Percentage of the population with access to safe drinking water.

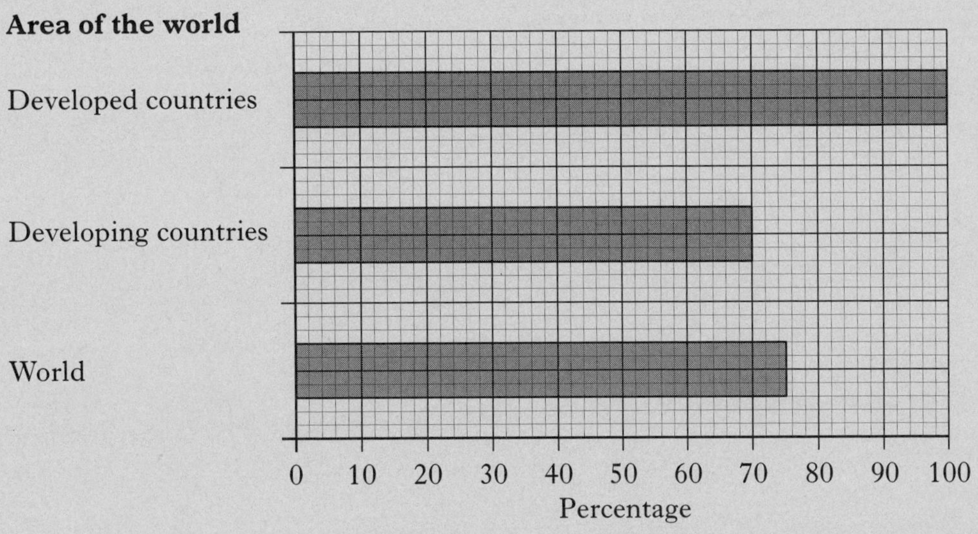

[Question 12 continues on *Pages twenty-four* and *twenty-five*

DO NOT WRITE IN THIS MARGIN

12. (continued) *Marks*

(*a*) Using the information from **Figure 1**, state the cause of death most similar in the developed and developing countries.

____ 1

(*b*) (i) Using the information from **Figure 1**, state the cause of death which differs most between developed and developing countries.

____ 1

(ii) Suggest an explanation for this difference.

____

____ 1

(*c*) Using the information from **Figure 1**, calculate the number of deaths which resulted from circulatory diseases in the developing countries.

________________________ 1

(*d*) With reference to **Figure 2**, which of the following results in the greatest percentage of child deaths?

Cause of death	*Tick*
Respiratory diseases	
Infections and parasitic diseases	
Birth related causes	
Other causes	

1

(*e*) (i) Using the information about **developing countries** in **Figures 1** and **2**, indicate which causes of death do not significantly affect children under the age of three.

____ 1

(ii) Suggest an explanation for this.

____

____ 1

[Question 12 continues on *Page twenty-five*

DO NOT WRITE IN THIS MARGIN

Marks

12. (continued)

(*f*) **Figure 2** shows that the incidence of child death due to malnutrition is low in developing countries. Suggest a reason for this. **1**

(*g*) Using the data in **Figure 3**, suggest an explanation for the incidence of diarrhoea as a cause of child deaths in the developing countries. **1**

(*h*) **Figure 3** shows that the percentage of the world population with access to safe drinking water is much closer to the percentage for developing countries than that of developed countries.

With reference to **Figure 1**, suggest a reason for this. **1**

[Section C begins on *Page twenty-six*

DO NOT WRITE IN THIS MARGIN

SECTION C

Marks

Answer BOTH questions 1 and 2 on the blank pages provided.

You may use labelled diagrams where appropriate.

1. Answer **either** A **or** B.

A. Give an account of the circulation of the blood, under the following headings:

(i) Pulmonary circulation (through heart and lungs); **5**

(ii) Systemic circulation (through heart and rest of the body). **5**

(10)

OR

B. Give an account of filtration and reabsorption in the kidney, under the following headings:

(i) Bowman's capsule and glomerulus; **4**

(ii) Kidney tubules. **6**

(10)

2. Answer **either** A **or** B.

A. Give an account of the roles of lipids in the body. **(10)**

OR

B. Describe how human activities disrupt the carbon cycle. **(10)**

[*END OF QUESTION PAPER*]

DO NOT WRITE IN THIS MARGIN

Marks

SPACE FOR ANSWERS

ADDITIONAL GRAPH PAPER FOR QUESTION 11(*c*)(i)

DO NOT WRITE IN THIS MARGIN

SPACE FOR ANSWERS

Marks

2001 HIGHER

FOR OFFICIAL USE

Total for Sections B & C

X009/301

NATIONAL QUALIFICATIONS 2001

MONDAY, 21 MAY
9.00 AM – 11.30 AM

HUMAN BIOLOGY HIGHER

Fill in these boxes and read what is printed below.

Full name of centre

Town

Forename(s)

Surname

Date of birth
Day Month Year

Scottish candidate number

Number of seat

SECTION A—Questions 1–30

Instructions for completion of Section A are given on page two.

SECTIONS B AND C

1 (a) All questions should be attempted.

(b) It should be noted that in **Section C** questions 1 and 2 each contain a choice.

2 The questions may be answered in any order but all answers are to be written in the spaces provided in this answer book, and must be written clearly and legibly in ink.

3 Additional space for answers and rough work will be found at the end of the book. If further space is required, supplementary sheets may be obtained from the invigilator and should be inserted inside the **front** cover of this book.

4 The numbers of questions must be clearly inserted with any answers written in the additional space.

5 Rough work, if any should be necessary, should be written in this book and then scored through when the fair copy has been written.

6 Before leaving the examination room you must give this book to the invigilator. If you do not, you may lose all the marks for this paper.

SCOTTISH QUALIFICATIONS AUTHORITY

SECTION A

Read carefully

1 Check that the answer sheet provided is for Human Biology Higher (Section A).

2 Fill in the details required on the answer sheet.

3 In this section a question is answered by indicating the choice A, B, C or D by a stroke made in **ink** in the appropriate place in the answer sheet—see the sample question below.

4 For each question there is only **one** correct answer.

5 Rough working, if required, should be done only on this question paper—or on the rough working sheet provided—**not** on the answer sheet.

6 At the end of the examination the answer sheet for Section A **must** be placed **inside** this answer book.

Sample Question

The digestive enzyme pepsin is most active in the

A mouth

B stomach

C duodenum

D pancreas.

The correct answer is **B**—stomach. A **heavy** vertical line should be drawn joining the two dots in the appropriate box in the column headed **B** as shown in the example on the answer sheet.

If, after you have recorded your answer, you decide that you have made an error and wish to make a change, you should cancel the original answer and put a vertical stroke in the box you now consider to be correct. Thus, if you want to change an answer D to an answer B, your answer sheet would look like this:

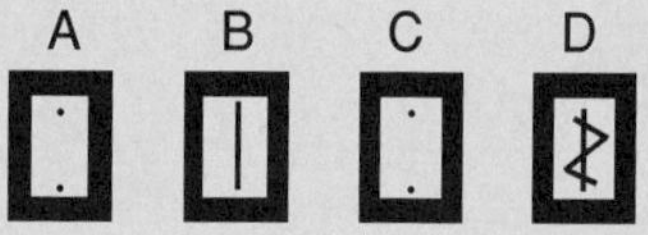

If you want to change back to an answer which has already been scored out, you should enter a tick (✓) to the **right** of the box of your choice, thus:

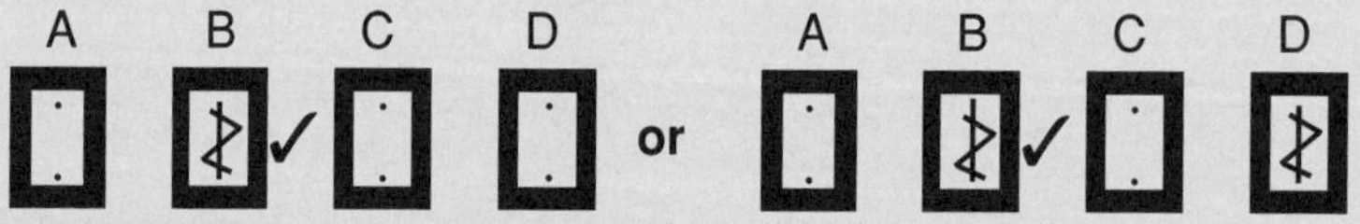

SECTION A

All questions in this section should be attempted.

Answers should be given on the separate answer sheet provided.

1. In respiration, the sequence of reactions resulting in the conversion of glucose to pyruvic acid is called

A the cytochrome system

B the TCA cycle

C the Krebs cycle

D glycolysis.

2. The diagram shows part of a liver cell with four parts labelled. In which part is most ATP produced?

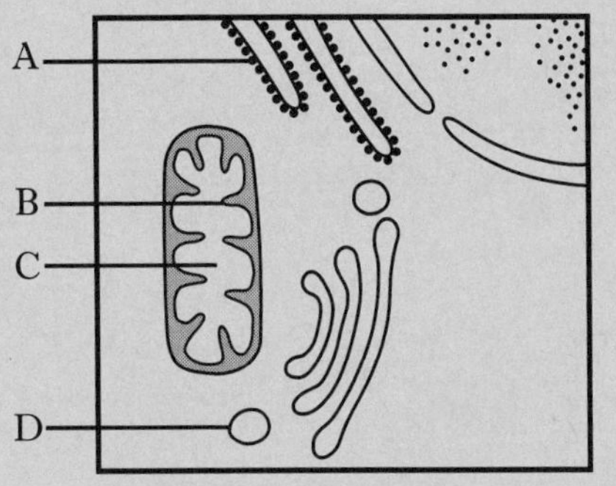

3. A DNA nucleotide could be formed from a molecule of phosphate together with

A ribose sugar and guanine

B ribose sugar and uracil

C deoxyribose sugar and guanine

D deoxyribose sugar and uracil.

4. If a DNA molecule contains 8000 nucleotides of which 20% are adenine, then the number of guanine nucleotides present is

A 1600

B 2000

C 2400

D 3200.

5. If the mass of DNA in a human liver cell is $6{\cdot}6 \times 10^{-12}$ g, the mass of DNA in a human sperm is likely to be

A $3{\cdot}3 \times 10^{-6}$ g

B $3{\cdot}3 \times 10^{-12}$ g

C $6{\cdot}6 \times 10^{-6}$ g

D $6{\cdot}6 \times 10^{-12}$ g.

6. A section of DNA has the following base sequence.

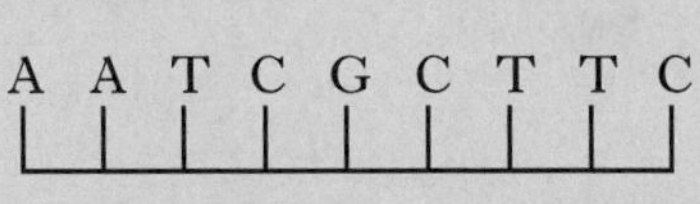

Identify the anti-codons of the three tRNA molecules which would align with the mRNA molecule transcribed from this section of DNA.

A AAU CGC UUC

B AAT CGC TTC

C TTA GCG AAG

D UUA GCG AAG

7. The cell organelle shown is magnified ten thousand times.

What is the actual size of the organelle?

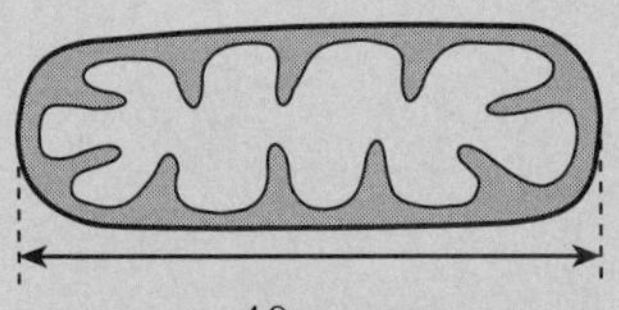

A 0·04 µm

B 0·4 µm

C 4 µm

D 40 µm

[Turn over

Questions 8 and 9 refer to the key shown below, used for the identification of carbohydrates.

1 { soluble 2
  insoluble ....glycogen

2 { Benedict's test positive.....................3
  Benedict's test negative....................sucrose

3 { Barfoed's test positive.......................4
  Barfoed's test negative ....lactose

4 { Clinistix test positive ....glucose
  Clinistix test negative ....fructose

8. Which line in the table of results below is **not** in agreement with the information contained in the key?

	Carbohydrate	*Clinistix test*	*Barfoed's test*	*Benedict's test*
A	sucrose	not tested	not tested	negative
B	glucose	positive	negative	positive
C	fructose	negative	positive	positive
D	lactose	not tested	negative	positive

9. Maltose is a soluble carbohydrate which gives a positive result with Benedict's but not with Barfoed's reagent. With which carbohydrate in the key could maltose be confused?

A Fructose

B Glucose

C Sucrose

D Lactose

10. The stages of infection of a host cell by a virus are listed below.

1 Host cell bursts, releasing new viruses.
2 Host cell DNA is inactivated.
3 Virus binds to host cell and injects DNA.
4 Virus DNA directs synthesis of new viruses.

The sequence in which these events occurs is

A 3,2,4,1

B 1,2,4,3

C 3,4,2,1

D 2,4,3,1.

Questions 11 and 12 refer to the information below.

The diagram shows the chromosome complement of cells during the development of abnormal sperm.

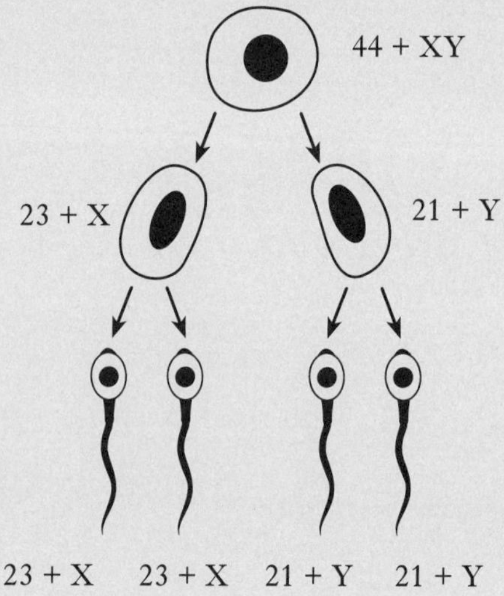

11. The diagram illustrates the effect of

A crossing over

B polygenic inheritance

C non-disjunction

D independent assortment of chromosomes.

12. A sperm with chromosome complement 23+X fertilises a normal haploid egg. What is the chromosome number and sex of the resulting zygote?

	Chromosome number	*Sex of zygote*
A	24	female
B	46	female
C	46	male
D	47	female

13. The colour of tooth enamel is a sex-linked characteristic. The allele for brown tooth enamel (e) is recessive to the allele for normal tooth enamel (E). The following family tree refers to this condition.

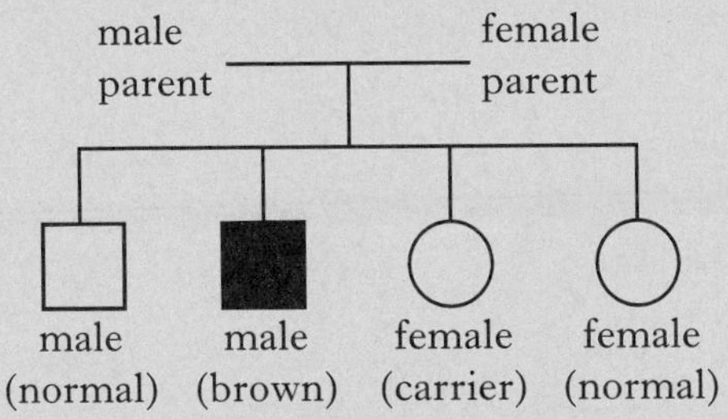

What are the genotypes of the parents?

A X^EY and X^EX^e

B X^EY and X^eX^e

C X^eY and X^EX^E

D X^eY and X^EX^e

Questions 14 and 15 refer to the following list of hormones.

A Follicle Stimulating Hormone (FSH)

B Luteinising Hormone (LH)

C Oestrogen

D Progesterone

14. Which hormone stimulates the production of testosterone by the testes?

15. Which hormone is produced by the corpus luteum?

16. Which of the following will **not** normally pass through the placenta between the mother and fetus?

A Oxygen

B Minerals

C Glucose

D Red blood cells

17. Fertility drugs may be used in the treatment of fertility to

A correct hormone imbalances

B reduce the pH of the oviduct

C stimulate mitosis in an egg

D protect sperm cells in the oviduct.

18. The durations of ventricular diastole and systole are shown below.

Diastole 0·4 seconds

Systole 0·2 seconds

What is the heart rate for this individual?

A 60 beats per minute

B 72 beats per minute

C 100 beats per minute

D 120 beats per minute

19. In which of the following pairs of tissues/organs, are red blood cells destroyed?

A Liver and lymph nodes

B Liver and spleen

C Bone marrow and duodenum

D Spleen and duodenum

20. Which of the following body fluids does **not** contain digestive enzymes?

A Saliva

B Gastric juice

C Pancreatic juice

D Bile

21. Which of the following results from an increase in the secretion of anti-diuretic hormone (ADH)?

A An increase in the permeability of the kidney tubules to water

B A decrease in the permeability of the kidney tubules to water

C An increase in the permeability of the glomerulus to water

D A decrease in the permeability of the glomerulus to water

22. Which of the following responses is caused by stimulation of the sympathetic nervous system?

A Increase in insulin production

B Increase in heart rate

C Increase in the flow of saliva

D Increase in peristalsis

23. A vertical section of the brain is shown in the diagram below.

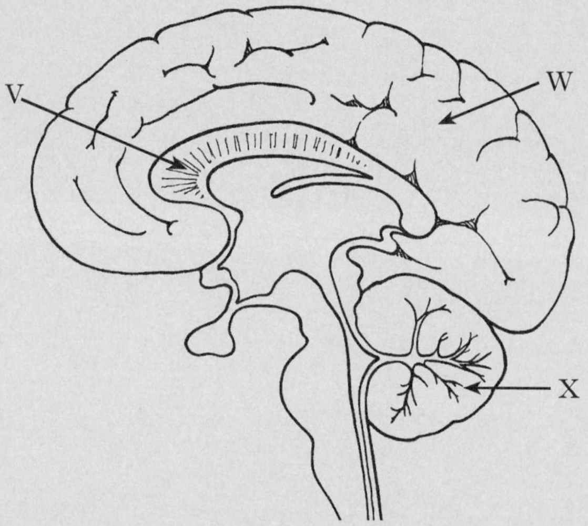

Which line of the table correctly labels the parts of the brain shown?

	V	*W*	*X*
A	corpus callosum	cerebellum	cerebrum
B	cerebellum	cerebrum	corpus callosum
C	corpus callosum	cerebrum	cerebellum
D	cerebrum	corpus callosum	cerebellum

24. Information is transferred between the two cerebral hemispheres by the

A corpus callosum

B medulla oblongata

C cerebellum

D hypothalamus.

25. A boy who is bitten by a large dog is subsequently frightened of all dogs. This behaviour pattern is an example of

A deindividuation

B extinction

C generalisation

D discrimination.

26. The development of phenotype is influenced by

A genetic factors only

B genetic factors and nutrition only

C the environment only

D the environment and genetic factors only.

27. The age structure of four different human populations is represented in the diagrams below. The bars indicate the relative numbers in each group.

Which diagram shows the population with greatest scope for growth?

A

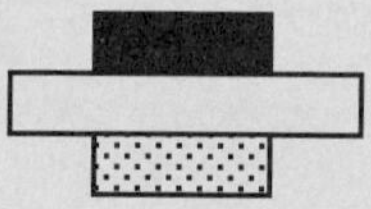

B

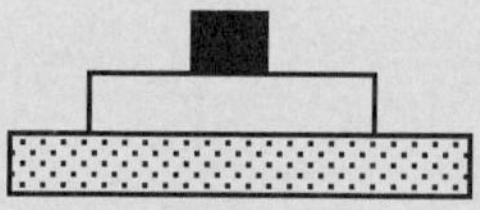

C

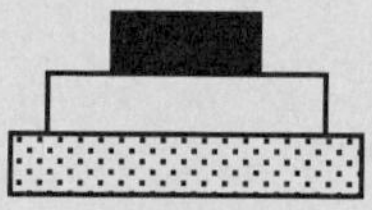

D

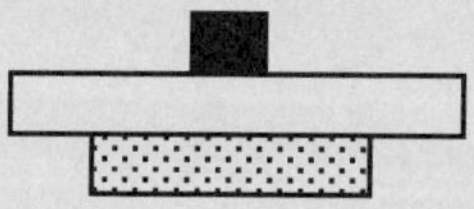

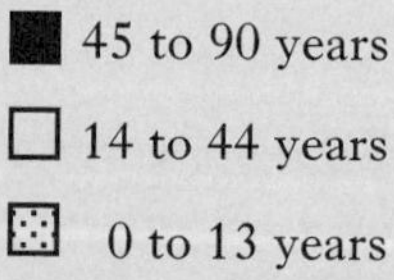

28. The interdependent biological and physical components in an area make up

A a habitat

B an ecosystem

C a food web

D a community.

29. What would be the effect of the discharge of raw sewage on the oxygen and nitrate concentrations of the water in a loch?

	Oxygen concentration	*Nitrate concentration*
A	increase	increase
B	increase	decrease
C	decrease	increase
D	decrease	decrease

30. The diagram shows a nitrogen cycle associated with the soil.

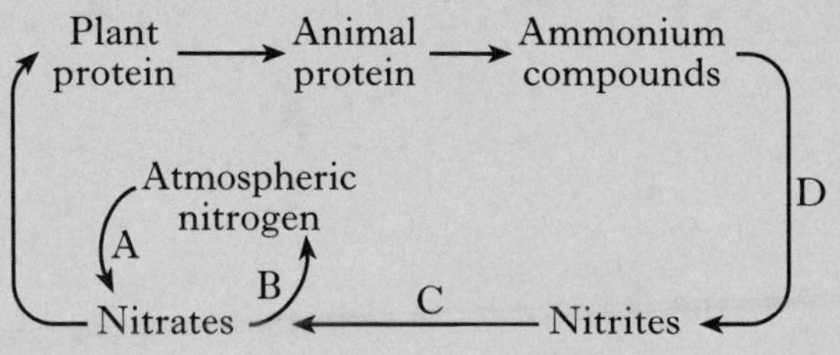

Which arrow indicates the activity of denitrifying bacteria?

Candidates are reminded that the answer sheet MUST be returned INSIDE this answer booklet.

[Turn over for Section B on *Page eight*

DO NOT WRITE IN THIS MARGIN

Marks

SECTION B

All questions in this section should be attempted.

1. The diagram below represents a reaction catalysed by an enzyme in the cytochrome system.

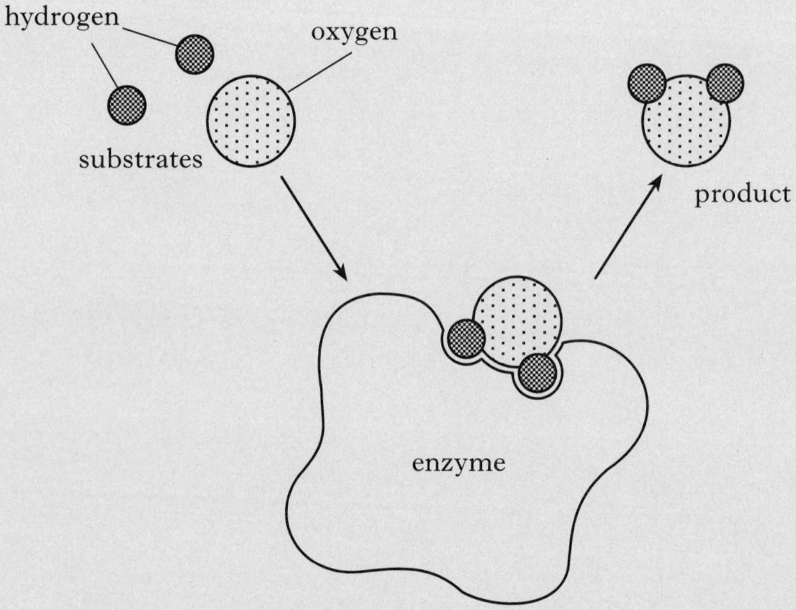

(*a*) (i) What name is given to the part of the enzyme where this reaction occurs?

________________________ **1**

(ii) In which organelle would this reaction take place?

________________________ **1**

(iii) Name the product of this reaction.

________________________ **1**

(*b*) Cyanide is a poison which inhibits this enzyme.

Suggest how cyanide is able to do this.

____

____ **1**

(*c*) Why do many enzyme-catalysed reactions require the presence of vitamins or minerals?

____ **1**

DO NOT WRITE IN THIS MARGIN

Marks

1. (continued)

(*d*) The graph shows the effect of increasing substrate concentration on the rate of this reaction.

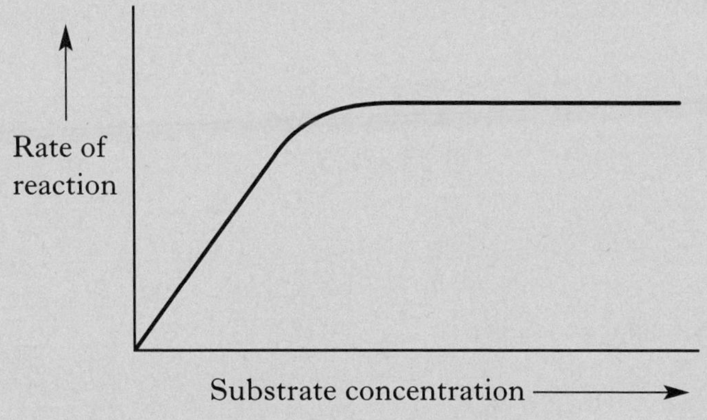

(i) Explain why the graph levels out at high substrate concentration.

____

______________________________ **1**

(ii) Assuming that the enzyme is operating at its optimum pH and temperature, suggest how the rate of reaction could be increased at high substrate concentrations.

____

______________________________ **1**

[Turn over

DO NOT WRITE IN THIS MARGIN

Marks

2. (*a*) The table below shows the relative concentrations of sodium and potassium ions in red blood cells and plasma.

	Sodium (units/litre)	*Potassium* (units/litre)
red blood cells	24	150
plasma	144	5

(i) Express, as simple ratios, the concentrations of sodium ions and potassium ions in the red blood cells and the plasma.

Space for calculation

Red blood cells : plasma

(1) sodium __________ : __________

(2) potassium __________ : __________ **1**

(ii) Suggest how the red blood cells maintain the potassium concentration gradient.

____

________________________________ **1**

(iii) When glucose is in short supply, the concentration of potassium in the red blood cells changes.

State whether the concentration will increase or decrease and give a reason for your answer.

Increase/decrease __________

Reason ________________________________

____

________________________________ **2**

DO NOT WRITE IN THIS MARGIN

Marks

2. (continued)

(*b*) Three samples of red blood cells were placed in different concentrations of sodium chloride solution for two minutes. The results of this treatment, when viewed under the microscope, are shown in the diagram below.

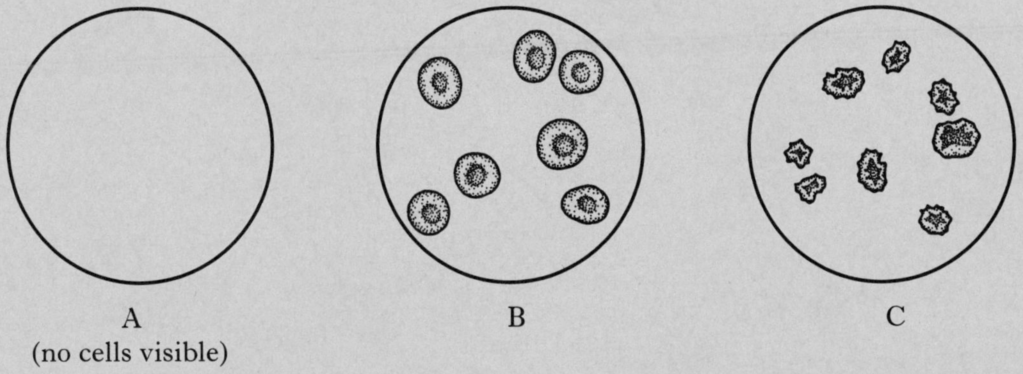

Using the information above, explain the appearance of the cells in each diagram.

A ____

____

B ____

____

C ____

____ **3**

[Turn over

DO NOT WRITE IN THIS MARGIN

Marks

3. Antigens on the surface of red blood cells enable different blood groups to be identified.

Four types of blood group are A, B, AB and O.

The diagram shows antigens on a red blood cell and antibodies in the surrounding plasma.

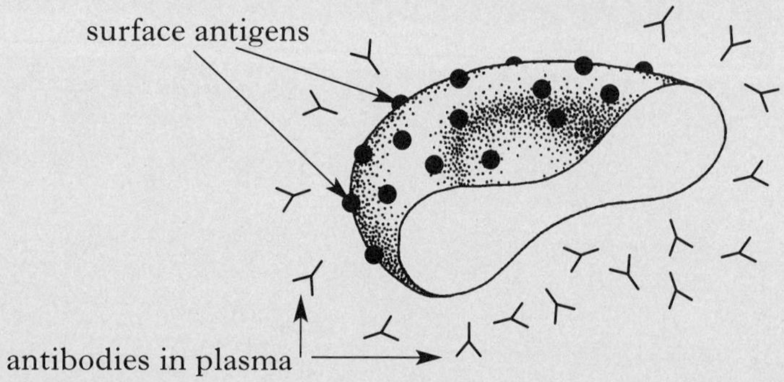

(*a*) Complete the table below to show the types of antigen and antibody present in individuals of each blood group.

Blood group	*Antigens present on surface of red blood cells*	*Antibodies present in plasma*
A	A	
B		anti-A
AB		
O		

2

(*b*) Which blood group(s) could be transfused safely into a person of blood group A?

________________ 1

DO NOT WRITE IN THIS MARGIN

Marks

3. (continued)

(*c*) The gene for this blood group has three alleles. Alleles A and B are co-dominant to allele O.

A man, heterozygous for blood group A, and a woman of blood group AB, have children.

(i) State the genotypes of the parents.

female ____________________ male ____________________ **1**

(ii) Complete the Punnett square below to show the genotypes of their gametes and the genotypes of the children they may have.

female gametes / male gametes		

2

(iii) What is the percentage chance that a child of these parents would have blood group A?

____________________ % **1**

[Turn over

DO NOT WRITE IN THIS MARGIN

Marks

4. The diagram below illustrates the first stage of the process of amniocentesis. The fluid removed from the uterus contains fetal cells which can be grown, stained and examined.

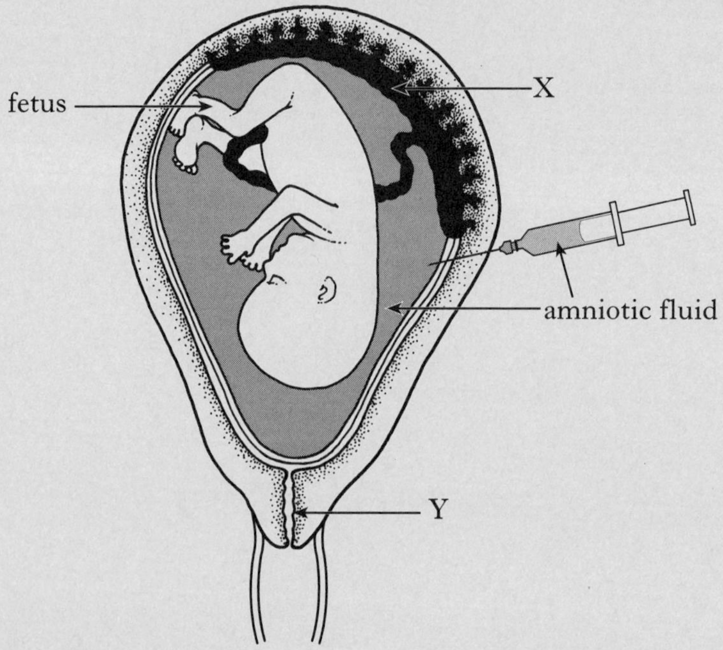

(*a*) (i) Identify structure X.

____________________________ **1**

(ii) Name a hormone produced by structure X during pregnancy.

____________________________ **1**

(*b*) (i) Identify structure Y.

____________________________ **1**

(ii) Structure Y produces mucus. What change occurs to this mucus during the fertile period?

____ **1**

DO NOT WRITE IN THIS MARGIN

Marks

4. (continued)

(*c*) The diagram represents a photograph of stained chromosomes from a fetal cell.

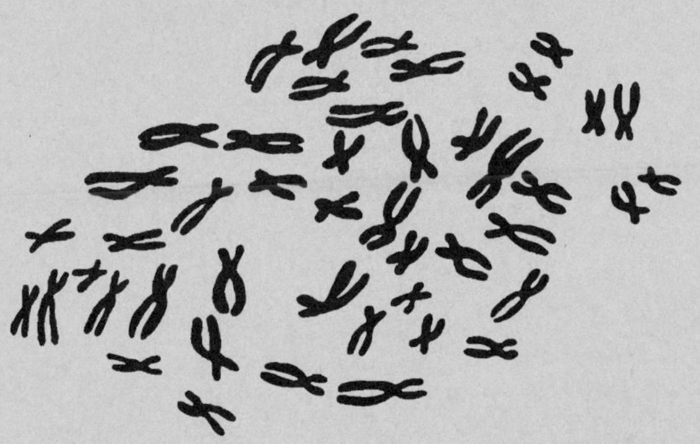

(i) The chromosomes may be cut out and arranged in homologous pairs.

Give **two** features of chromosomes which allow homologous pairs to be identified.

1 ______________________________

2 ______________________________ **1**

(ii) What name is given to a set of chromosomes arranged in pairs?

______________________________ **1**

(iii) How could the sex of the fetus be identified from this paired arrangement?

____

______________________________ **1**

(iv) What other information could be obtained which would be of value in pre-natal screening?

____

______________________________ **1**

[Turn over

DO NOT WRITE IN THIS MARGIN

Marks

5. The diagram below shows the parts of a kidney nephron involved in ultrafiltration.

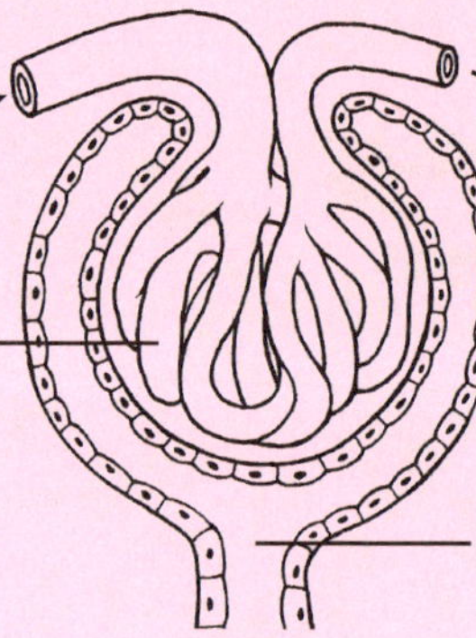

(*a*) (i) Part A consists of a bundle of capillaries. Name part A.

______________________________ **1**

(ii) What feature, shown in the diagram, results in high blood pressure within structure A?

____ **1**

(iii) Into which part of the nephron does the filtrate flow immediately after leaving the Bowman's capsule?

______________________________ **1**

(*b*) The table shows the composition of filtrate and urine.

Substance	*Mass in filtrate* (g/day)	*Mass in urine* (g/day)
Sodium ions	600	6
Potassium ions	35	2
Glucose	200	0
Urea	60	36
Water	180 000	1500

(i) Name the process which results in the differences between filtrate and urine, shown in the table.

______________________________ **1**

(ii) What percentage of urea returns to the blood as the filtrate flows through the nephron?

_______________ % **1**

(iii) Predict how the composition of urine would differ if the individual was an untreated diabetic.

____ **1**

DO NOT WRITE IN THIS MARGIN

Marks

6. The diagram below shows the parasympathetic nerve which runs between the central nervous system and the heart.

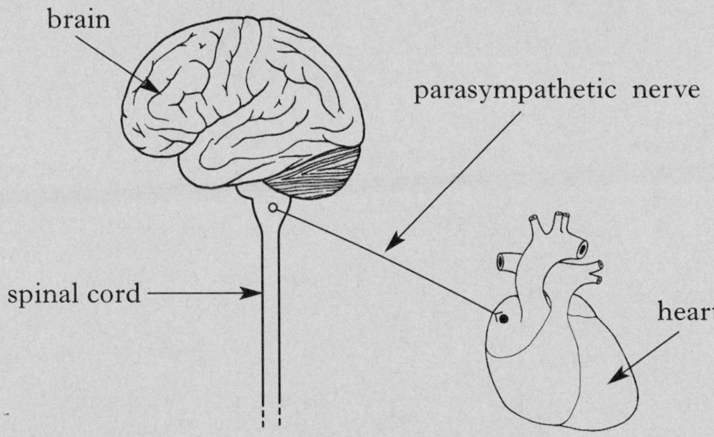

(*a*) (i) Which subdivision of the peripheral nervous system contains parasympathetic nerves?

______________________________ **1**

(ii) In which part of the brain does this parasympathetic nerve originate?

______________________________ **1**

(*b*) (i) Name the part of the right atrium which is stimulated by the parasympathetic nerve.

______________________________ **1**

(ii) State the effect of parasympathetic stimulation on the heart.

____ **1**

(*c*) Describe another effect which the parasympathetic nervous system has on the body.

____

____ **1**

[Turn over

7. In humans, alterations in the level of exercise bring about changes in pulse rate, stroke volume and ventilation rate. The level of exercise is measured as rate of oxygen uptake.

Graph 1 gives information about the heart. It shows how pulse rate and stroke volume change with the level of exercise in an individual.

Stroke volume is the volume of blood pumped from the heart in one beat.

Graph 1

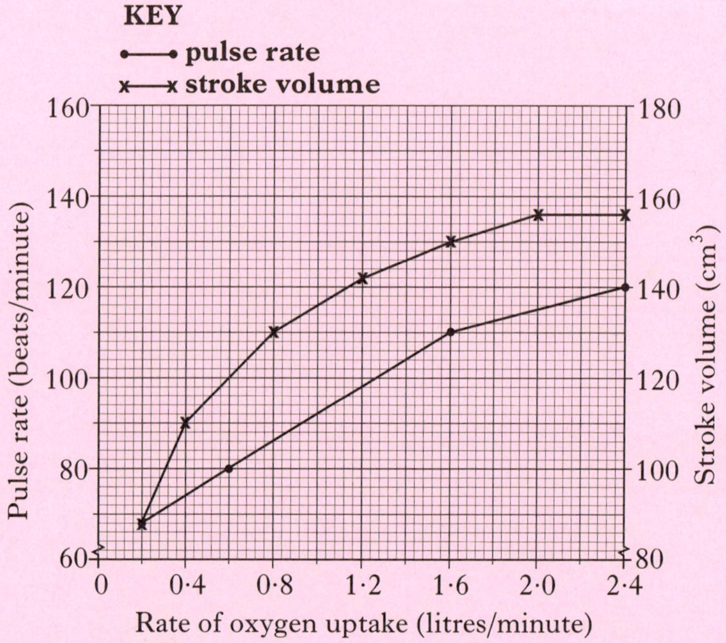

Graph 2 gives information about the lungs. It shows how the ventilation rate changes with the level of exercise in the same individual.

Ventilation rate is the volume of air inhaled during one minute.

Graph 2

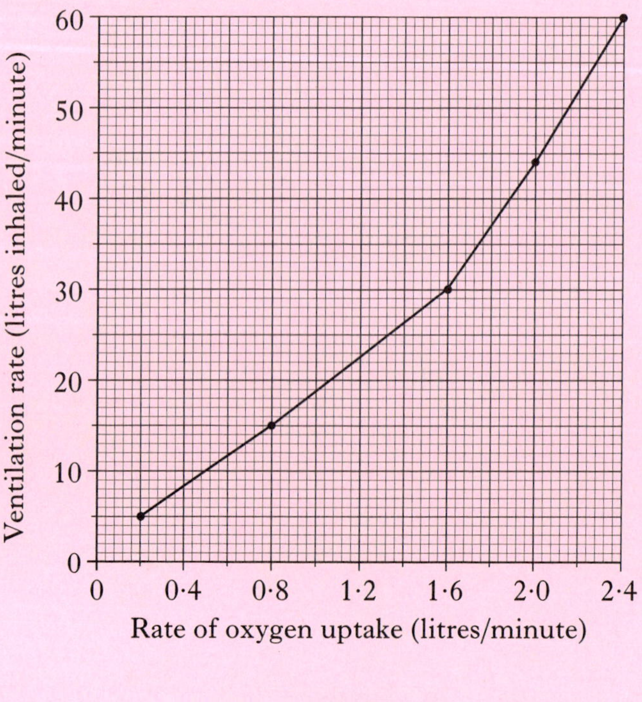

DO NOT WRITE IN THIS MARGIN

Marks

7. (continued)

(*a*) (i) With reference to **Graph 1**, what is the pulse rate and the stroke volume when the rate of oxygen uptake is 0·8 litres/minute?

Pulse rate ________________ Stroke volume ________________ **2**

(ii) What is the stroke volume when the pulse rate is 74 beats per minute?

____________ cm^3 **1**

(iii) What is the total volume of blood leaving the heart in one minute when the rate of oxygen uptake is 1·6 litres/minute?

Space for calculation

________________ litres/minute **1**

(iv) From **Graph 1**, compare the pattern of changes in pulse rate and stroke volume as oxygen uptake increases.

____

____

____ **2**

(*b*) (i) Fresh air contains 20% oxygen. From **Graph 2**, what is the volume of oxygen inhaled per minute when the rate of oxygen uptake is 1·6 litres/minute?

Space for calculation

________________ litres **1**

(ii) What additional information would be required to calculate the average volume of air taken in during each breath at any time?

____ **1**

(*c*) (i) With reference to **both graphs**, state the ventilation rate when this individual's pulse rate is 100 beats per minute.

____________ litres/minute **1**

(ii) Complete the table below by ticking the correct statement(s).

Statement	*Tick* (✓)
The rate at which pulse rate changes is highest at low rates of oxygen uptake.	
When ventilation rate doubles, the rate of oxygen uptake doubles.	

1

(*d*) Name the blood vessel which carries deoxygenated blood from the heart to the lungs.

____________________________ **1**

DO NOT WRITE IN THIS MARGIN

Marks

8. Part of a neurone is shown in the diagram below.

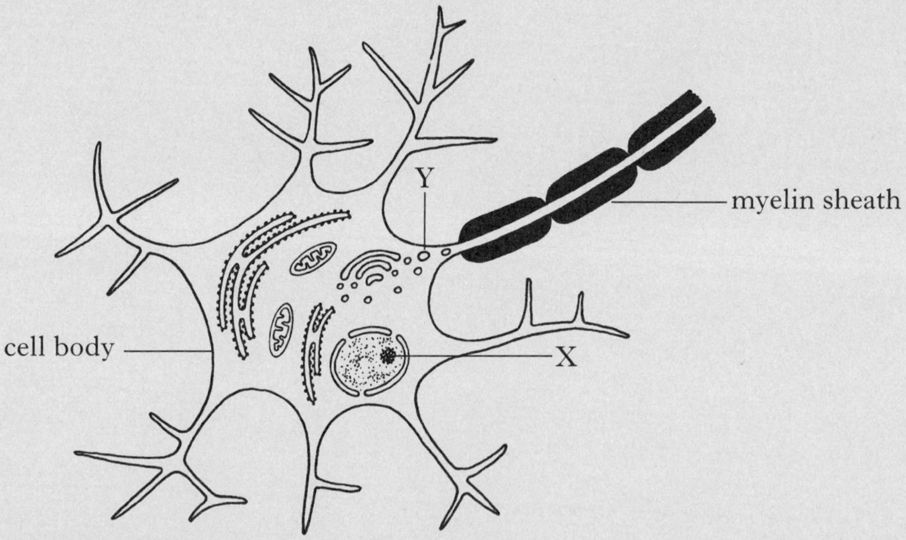

(*a*) State whether the neurone shown is a sensory or motor neurone and give a reason for your answer.

Type of neurone ______________________________

Reason for answer ____

____ **1**

(*b*) Name structure X and state its function.

Name ____________________________

Function ____ **2**

(*c*) (i) Name structure Y.

____________________________ **1**

(ii) Similar structures are found in the synaptic knob. What do they contain?

____________________________ **1**

DO NOT WRITE IN THIS MARGIN

8. (continued) *Marks*

(*d*) In the disorder Multiple Sclerosis, the myelin sheath is damaged by the body's own defence system.

(i) What effect does this have on the function of the nerve fibre?

____

____ **1**

(ii) What term is used to describe a disorder where the body's defence system destroys its own cells?

______________________ **1**

(*e*) Draw an arrow on the diagram to show the direction of an impulse in a dendrite. **1**

(*f*) Diverging neural pathways always contain the type of neurone shown opposite. Explain how diverging pathways allow humans to perform a task such as threading a needle.

____

____

____

____ **2**

[Turn over

DO NOT WRITE IN THIS MARGIN

Marks

9. "Fast foods" are now very much part of the culture of the developed world.

The table below gives information about a beef burger.

Beef burger

Nutritional analysis/100 g	
Energy	1500 kJ
Protein	12 g
Carbohydrate	8 g
Fat	30 g
Fibre	1 g
Sodium	1 g

(*a*) A boy ate little else but beef burgers every day.

With reference to the table, explain why the boy might suffer from malnutrition but not starvation.

____

____

____ **2**

(*b*) Increased demand for cheap beef has had an impact on the natural ecosystems of developing countries.

Suggest how this demand affects natural ecosystems and local water supplies.

Natural ecosystems ______________________________

____

Local water supplies ______________________________

____ **2**

(*c*) Why is the production of beef an inefficient use of land in a developing country where there is a large population to feed?

____

____ **1**

DO NOT WRITE IN THIS MARGIN

Marks

9. (continued)

(*d*) The carbohydrate in the burger comes from wheat.

Modern varieties of wheat have been produced by selective breeding.

Describe an improvement brought about by selective breeding of crop plants such as wheat.

____

____

____ **1**

(*e*) (i) Pesticides are frequently applied to growing crops. What is a pesticide?

____ **1**

(ii) Describe **one** advantage of using pesticides.

____

____

____ **1**

(*f*) Genetic manipulation is now used to produce new varieties of organisms.

Describe **one** advantage of this technique compared to selective breeding.

____

____

____ **1**

[Turn over

Marks

10. The apparatus shown in **Figure 1** was used to investigate the effect of nitrates on the growth of grass. Grass seedlings were grown in seven different culture solutions. The experiment was repeated ten times.

Figure 2 shows the nitrate concentrations of the culture solutions and the results of the experiment.

Figure 1

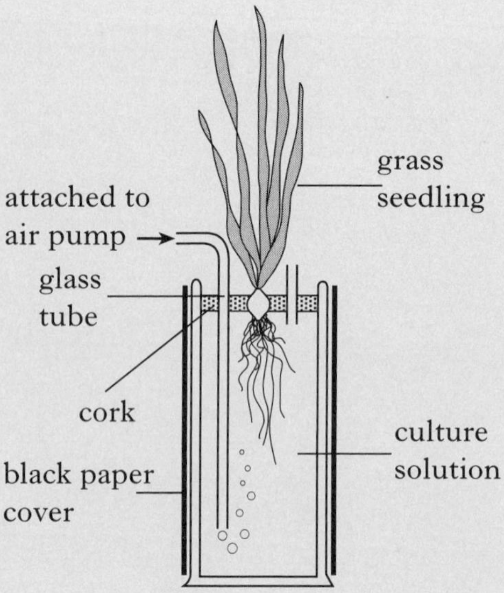

Figure 2

Culture solution	*Nitrate content* (g/litre)	*Average height of plants after 6 weeks of growth* (cm)
A	0	3
B	0·5	12
C	1·0	17
D	1·5	23
E	2·0	25
F	2·5	24
G	3·0	25

(*a*) Plot a line graph to illustrate the experimental results.

(Additional graph paper, if required, will be found on page 28.)

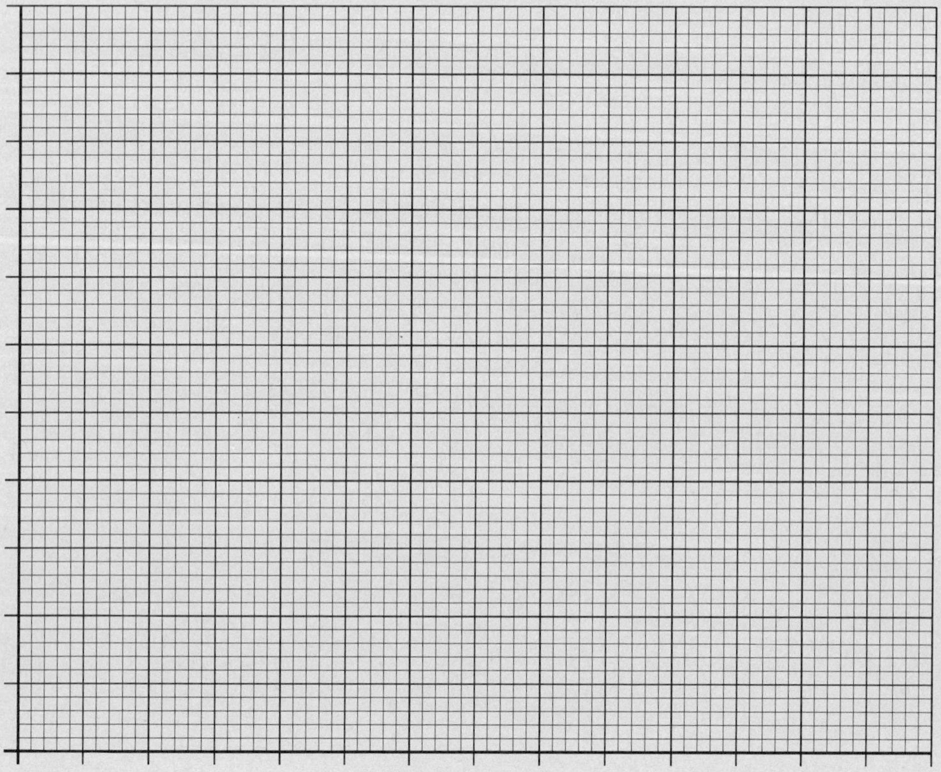

3

DO NOT WRITE IN THIS MARGIN

10. (continued)

Marks

(*b*) Name **two** variables which should be controlled in this experiment.

1 ______________________________

2 ______________________________ **1**

(*c*) During the experiment air was pumped into the jars through the glass tubes.

Suggest why this was necessary.

____

______________________________ **1**

(*d*) What feature of the experiment makes the results more reliable?

____

______________________________ **1**

(*e*) State **one** other feature of grass plants which could be observed or measured to assess the effects of nitrate.

______________________________ **1**

(*f*) Predict how the results of this experiment would be different in culture solutions A, B and C if clover or pea plants had been used instead of grass plants. Give a reason for your answer.

Prediction ______________________________ **1**

Reason ______________________________ **1**

(*g*) A farmer wishes to purchase nitrate fertiliser. In what way could the information from this experiment be useful to the farmer?

____

______________________________ **1**

[Section C begins on *Page twenty-six*

DO NOT WRITE IN THIS MARGIN

Marks

SECTION C

Both questions in this section should be attempted.

Note that each question contains a choice.

Questions 1 and 2 should be attempted on the blank pages which follow.

Supplementary sheets, if required, may be obtained from the invigilator.

Labelled diagrams may be used where appropriate.

1. Answer **either** A **or** B.

A. Give an account of memory under the following headings:

(i) encoding; **6**

(ii) storage; **2**

(iii) retrieval. **2**

(10)

OR

B. Give an account of immunisation under the following headings:

(i) artificial active immunity; **6**

(ii) artificial passive immunity; **2**

(iii) the impact of vaccination on childhood diseases. **2**

(10)

In question 2 ONE mark is available for coherence and ONE mark is available for relevance.

2. Answer **either** A **or** B.

A. Describe the biological basis of contraception. **(10)**

OR

B. Outline the involuntary mechanisms involved in temperature control. **(10)**

[*END OF QUESTION PAPER*]

DO NOT WRITE IN THIS MARGIN

Marks

SPACE FOR ANSWERS

DO NOT WRITE IN THIS MARGIN

Marks

SPACE FOR ANSWERS

ADDITIONAL GRAPH PAPER FOR QUESTION 10(*a*)

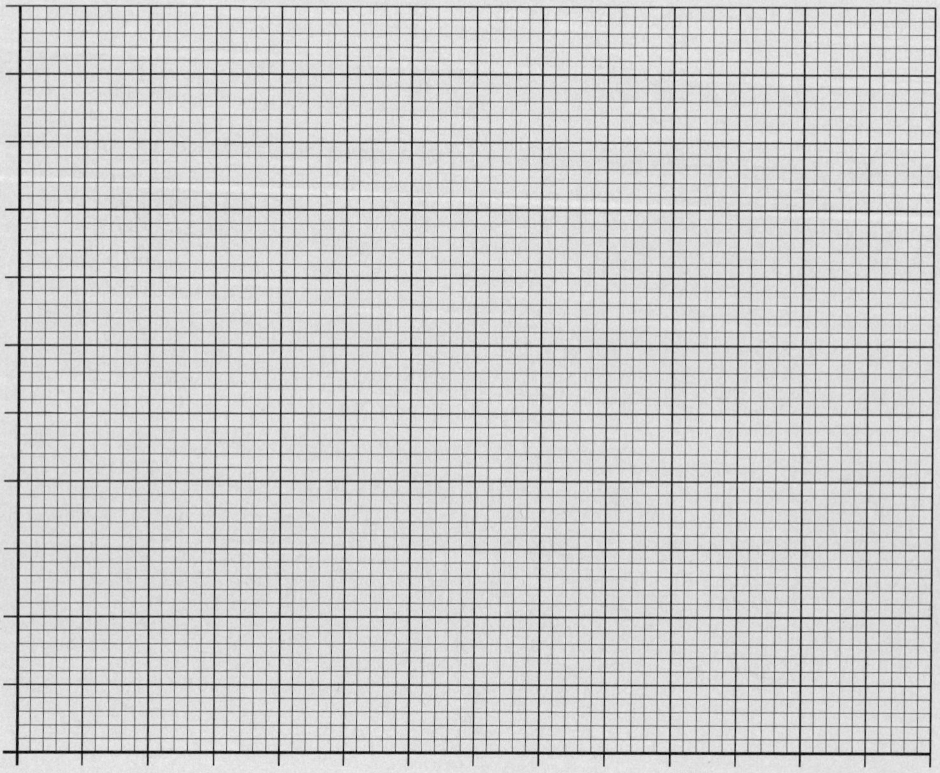

2002 HIGHER

FOR OFFICIAL USE

Total for Sections B & C

X009/301

NATIONAL QUALIFICATIONS 2002

FRIDAY, 31 MAY
1.00 PM – 3.30 PM

HUMAN BIOLOGY HIGHER

Fill in these boxes and read what is printed below.

Full name of centre

Town

Forename(s)

Surname

Date of birth
Day Month Year

Scottish candidate number

Number of seat

SECTION A—Questions 1–30

Instructions for completion of Section A are given on page two.

SECTIONS B AND C

1 (a) All questions should be attempted.
  (b) It should be noted that in **Section C** questions 1 and 2 each contain a choice.

2 The questions may be answered in any order but all answers are to be written in the spaces provided in this answer book, and must be written clearly and legibly in ink.

3 Additional space for answers and rough work will be found at the end of the book. If further space is required, supplementary sheets may be obtained from the invigilator and should be inserted inside the **front** cover of this book.

4 The numbers of questions must be clearly inserted with any answers written in the additional space.

5 Rough work, if any should be necessary, should be written in this book and then scored through when the fair copy has been written.

6 Before leaving the examination room you must give this book to the invigilator. If you do not, you may lose all the marks for this paper.

SCOTTISH QUALIFICATIONS AUTHORITY

SECTION A

Read carefully

1 Check that the answer sheet provided is for Human Biology Higher (Section A).

2 Fill in the details required on the answer sheet.

3 In this section a question is answered by indicating the choice A, B, C or D by a stroke made in **ink** in the appropriate place in the answer sheet—see the sample question below.

4 For each question there is only **one** correct answer.

5 Rough working, if required, should be done only on this question paper—or on the rough working sheet provided—**not** on the answer sheet.

6 At the end of the examination the answer sheet for Section A **must** be placed **inside** this answer book.

Sample Question

The digestive enzyme pepsin is most active in the

A mouth

B stomach

C duodenum

D pancreas.

The correct answer is **B**—stomach. A **heavy** vertical line should be drawn joining the two dots in the appropriate box in the column headed **B** as shown in the example on the answer sheet.

If, after you have recorded your answer, you decide that you have made an error and wish to make a change, you should cancel the original answer and put a vertical stroke in the box you now consider to be correct. Thus, if you want to change an answer D to an answer B, your answer sheet would look like this:

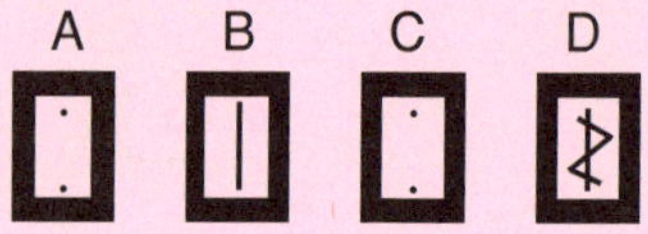

If you want to change back to an answer which has already been scored out, you should enter a tick (✓) to the **right** of the box of your choice, thus:

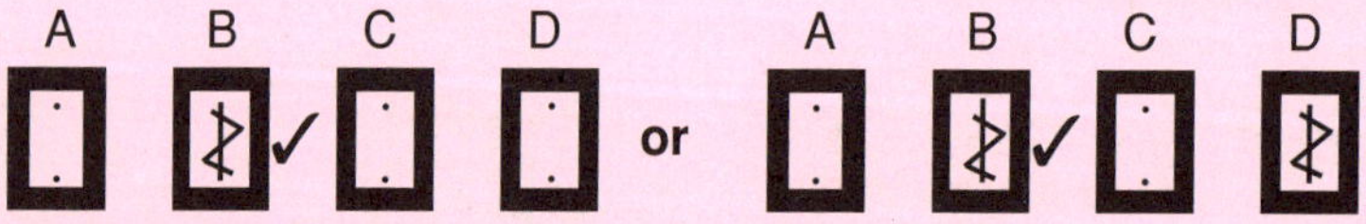

SECTION A

All questions in this section should be attempted.

Answers should be given on the separate answer sheet provided.

1. The cell membrane is chiefly composed of

A carbohydrates and lipids

B carbohydrates and proteins

C proteins and lipids

D proteins and nucleic acids.

2. Thirty percent of bases in a DNA molecule are adenine. The percentage of cytosine bases in the same molecule is

A 20%

B 30%

C 40%

D 70%.

3. Which of the following must be present for glycolysis to occur?

A Glucose and oxygen

B ATP and oxygen

C Glucose and ATP

D ATP and pyruvic acid

4. Glycolysis takes place in the

A nucleus

B cristae of mitochondria

C matrix of mitochondria

D cytoplasm.

5. The following diagram represents stages in the complete breakdown of glucose in aerobic respiration.

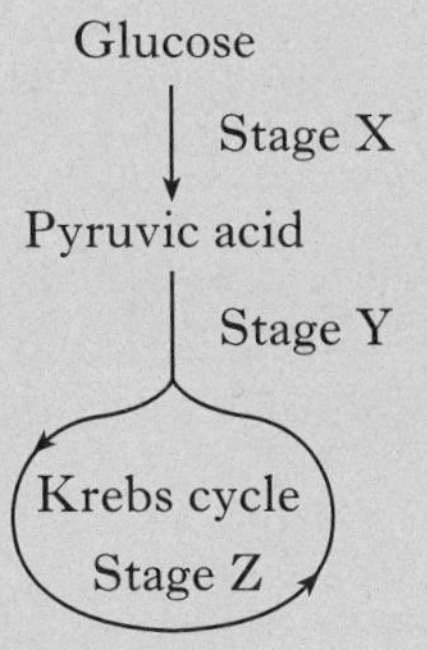

At which stage, or stages, is carbon dioxide released?

A Stages X and Z

B Stages X and Y

C Stages Y and Z

D Stage Z only

[Turn over

6. The diagram below shows part of a metabolic pathway. Each stage is controlled by an enzyme.

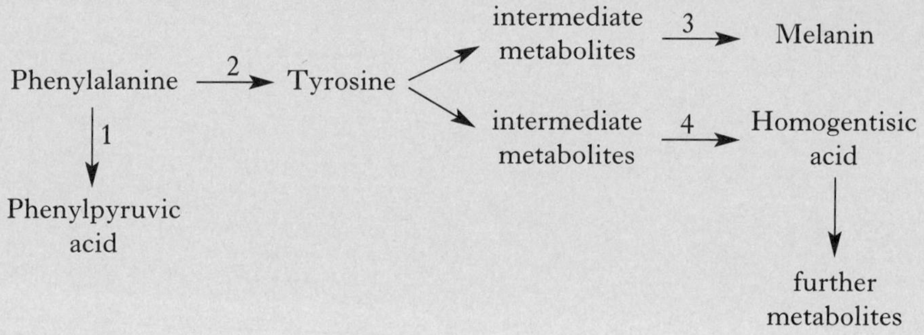

Phenylketonuria (PKU) is caused by a mutation of the gene required to make enzyme

A 1

B 2

C 3

D 4.

7. A stock solution has a concentration of 1 M. 100 cm^3 of a 0·6 M solution can be prepared by adding

A 40 cm^3 of stock solution to 60 cm^3 of water

B 60 cm^3 of stock solution to 40 cm^3 of water

C 60 cm^3 of stock solution to 100 cm^3 of water

D 100 cm^3 of stock solution to 60 cm^3 of water.

8. The table below shows some genotypes and phenotypes associated with forms of sickle-cell anaemia.

Genotype	*Phenotype*
Hb^AHb^A	normal
Hb^AHb^S	sickle-cell trait
Hb^SHb^S	acute sickle-cell anaemia

A normal man marries a woman with the sickle-cell trait. What are the chances that any child born to them will have acute sickle-cell anaemia?

A None

B 1 in 1

C 1 in 2

D 1 in 4

9. The diagram below shows the transmission of the gene for albinism.

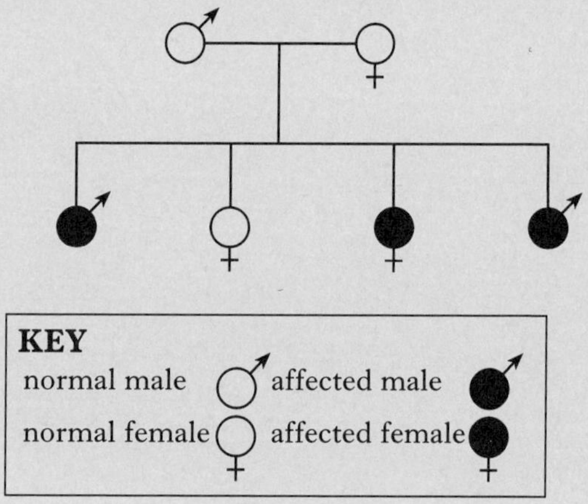

This condition is inherited as a characteristic which is

A dominant and not sex-linked

B recessive and not sex-linked

C dominant and sex-linked

D recessive and sex-linked.

10. Colour-blindness is a recessive, sex-linked characteristic controlled by the allele b.

Two parents with normal vision have a colour-blind boy.

The genotypes of the parents are

A X^BY and X^bX^b

B X^bY and X^BX^B

C X^bY and X^BX^b

D X^BY and X^BX^b.

11. The function of the interstitial cells in the human testes is to

A act as a store for sperm cells

B produce semen

C cause sperm cells to mature

D produce testosterone.

12. The diagram below shows a section through an ovary which contains developing eggs.

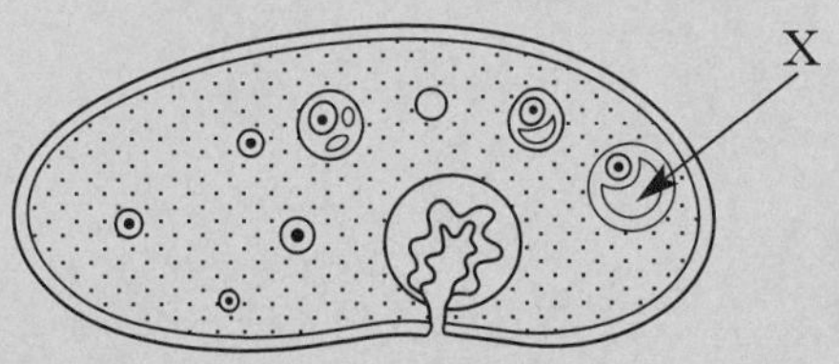

The structure labelled X is

A the endometrium

B a Graafian follicle

C the amnion

D a corpus luteum.

13. Which of the following materials are exchanged between maternal and fetal blood by diffusion?

A Oxygen and proteins

B Carbon dioxide and antibodies

C Oxygen and glucose

D Carbon dioxide and oxygen

14. The graph below represents an arterial blood pressure trace.

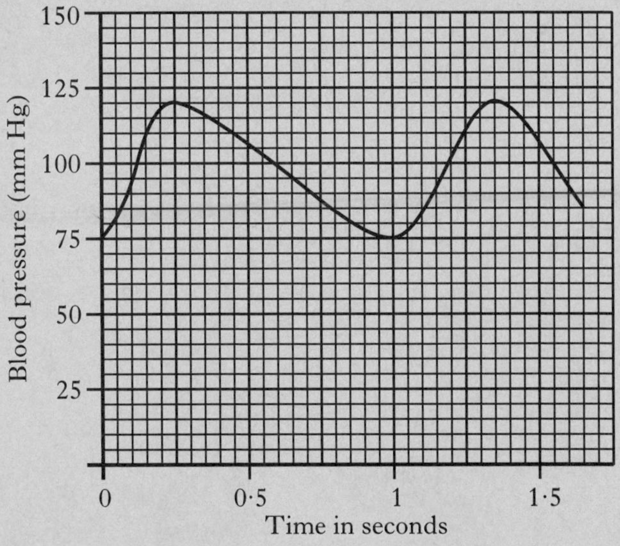

The blood pressure would be recorded as

A $\frac{120}{75}$

B $\frac{75}{120}$

C $\frac{104}{75}$

D $\frac{75}{104}$

15. The percentage distribution of blood groups in Scotland is shown below.

	Blood Group			
	O	A	B	AB
Scots (%)	51	35	11	3

What percentage of Scots could be given a blood transfusion of blood group A?

A 35%

B 38%

C 51%

D 86%

16. A person produces 1·5 litres of urine in 24 hours. The urine contains 36 g of urea.

What is the concentration of urea in the urine?

A $1{\cdot}0\ g/100\ cm^3$

B $2{\cdot}4\ g/litre$

C $2{\cdot}4\ g/100\ cm^3$

D $3{\cdot}6\ g/100\ cm^3$

17. Stimulation of the sympathetic nerves causes

A vasoconstriction of arterioles in the skin

B vasoconstriction of the coronary arterioles

C vasodilation of arterioles of the gut

D vasodilation of arterioles in the salivary glands.

18. The flow chart shows the sub-divisions of the human nervous system. Which letter represents the autonomic nervous system?

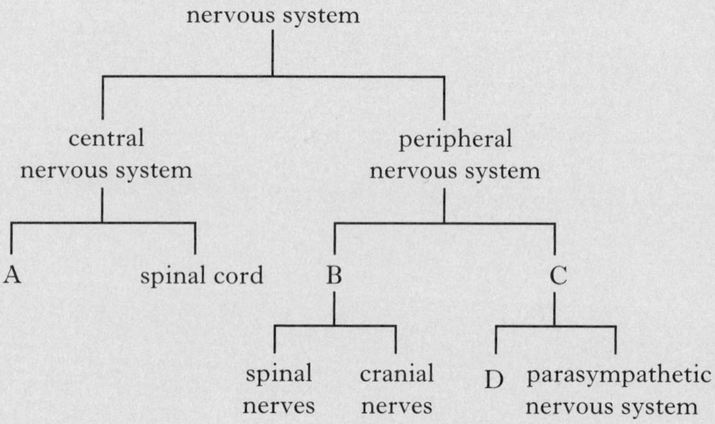

19. The somatic nervous system controls the

A skeletal muscles

B heart and blood vessels

C endocrine glands

D muscular wall of the gut.

20. The speed of impulse transmission along an axon is promoted by

A diffusion of neurotransmitters

B converging neural pathways

C diverging neural pathways

D myelination of fibres.

21. Which of the following statements about diverging neural pathways is correct?

A They accelerate the transmission of sensory impulses.

B They suppress the transmission of sensory impulses.

C They increase the degree of fine motor control.

D They decrease the degree of fine motor control.

22. The following diagram represents four neurones in a nervous pathway.

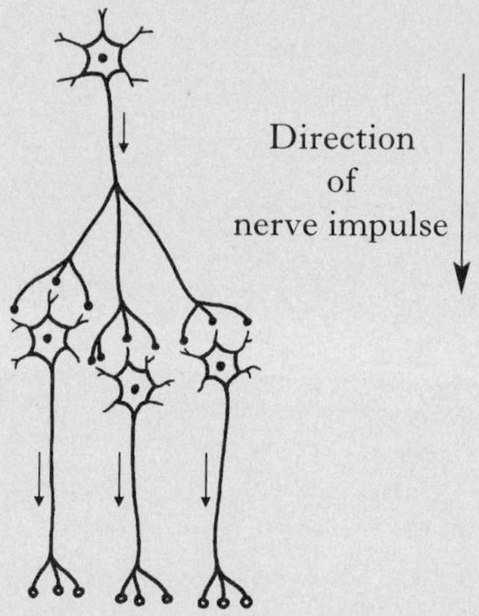

Which line of the table describes correctly the pathway?

	Type of Pathway	
A	sensory	convergent
B	motor	convergent
C	sensory	divergent
D	motor	divergent

23. The retrieval of information from long term memory is often aided by remembering the situation in which the information was encoded. This is described as using

A contextual cues

B chunking techniques

C rehearsal methods

D memory span.

24. A child, who was scratched by a black cat, now responds with a fear of all cats. This is an example of

A shaping

B reinforcement

C generalisation

D discrimination.

25. The diagram below shows the ages (in months) at which children reach various stages in their development.

The left end of each bar indicates the age by which 25% of infants have reached the stated performance.

The right end of each bar indicates the age by which 90% of infants have reached the stated performance.

The vertical bar indicates the age by which 50% of infants have reached the stated performance.

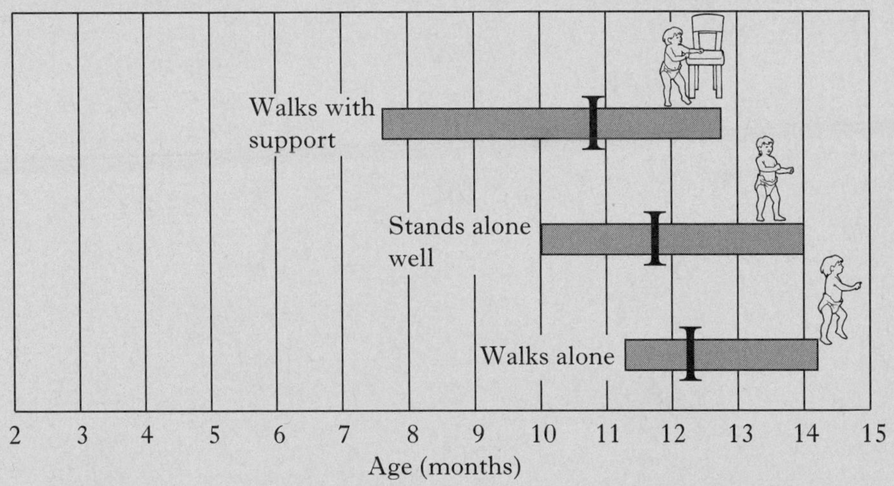

A nine-month-old infant can stand without support but cannot walk without support.

In what percentage of the population does this child lie?

A Less than 25%

B Around 25%

C Around 50%

D Greater than 50%

26. The diagram below shows the average lifespan of people in Britain between 1900 and 1990.

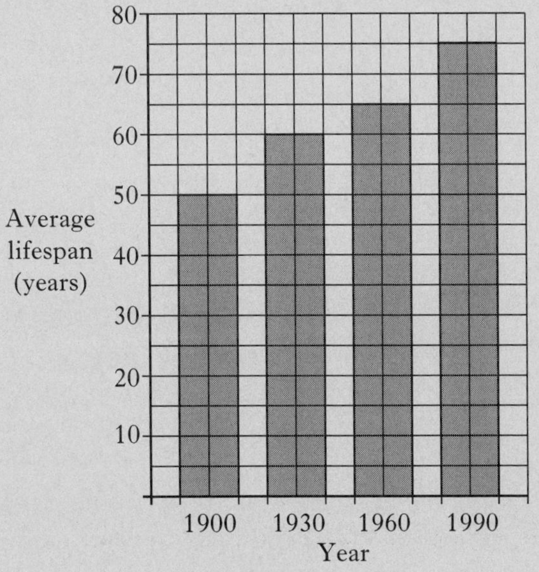

What is the percentage increase in lifespan during this period?

A 25%

B 45%

C 50%

D 75%

27. In the nitrogen cycle, which of the following processes is carried out by nitrifying bacteria?

The conversion of

A nitrate to ammonia

B ammonia to nitrate

C nitrogen gas to ammonia

D nitrogen gas to nitrate.

28. In a river, samples of water from above and below a sewage outlet were compared. Which comparison is correct?

	Sample above sewage outlet	*Sample below sewage outlet*
A	high oxygen concentration	many bacteria
B	high oxygen concentration	few bacteria
C	low oxygen concentration	many bacteria
D	low oxygen concentration	few bacteria

29. The graphs below contain information about the population of Britain.

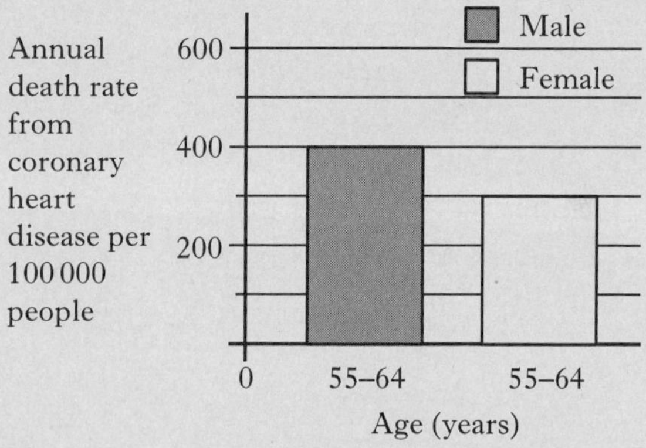

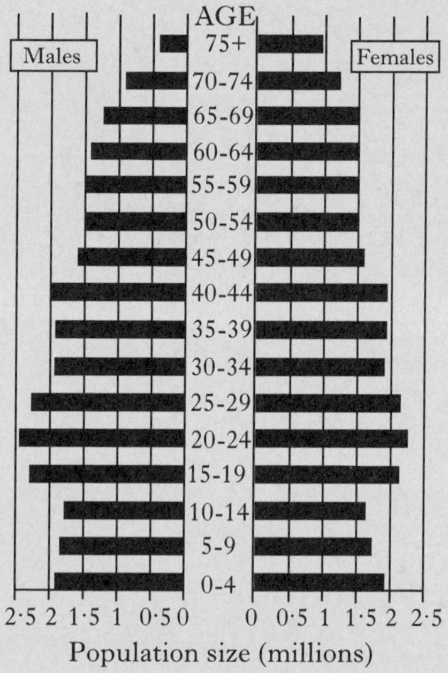

How many British women between 55 and 64 years of age die from coronary heart disease annually?

A 300

B 4500

C 9000

D 21 000

30. Which of the following pairs of gases are the principal contributors to the greenhouse effect?

A Nitrogen and carbon dioxide

B Carbon dioxide and methane

C Ammonia and carbon dioxide

D Nitrogen and methane

Candidates are reminded that the answer sheet MUST be returned INSIDE this answer booklet.

[Turn over for Section B]

DO NOT WRITE IN THIS MARGIN

Marks

SECTION B

All questions in this section should be attempted.

1. The diagram below shows a section through a nucleus and associated cell structures.

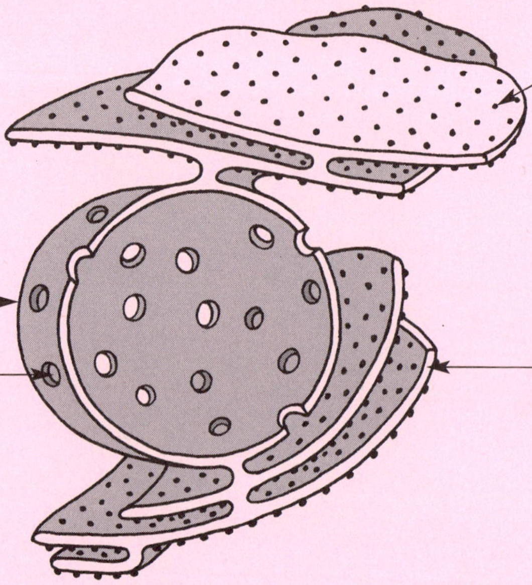

(*a*) (i) Name organelle **X**.

________________________ **1**

(ii) What type of substance is manufactured by organelle **X**?

________________________ **1**

(iii) Give an example of such a substance.

________________________ **1**

(*b*) (i) The structure labelled **Y** is composed of sheets of membranes.

What name is given to this structure?

________________________ **1**

(ii) Structure **Y** can transport substances to another organelle within the cell.

Give an example of such an organelle and state its function.

Example ________________________ **1**

Function ________________________ **1**

(*c*) Why is it necessary to have pores in the nuclear membrane?

____

____ **1**

DO NOT WRITE IN THIS MARGIN

Marks

2. The diagram shows the synthesis of a peptide chain.

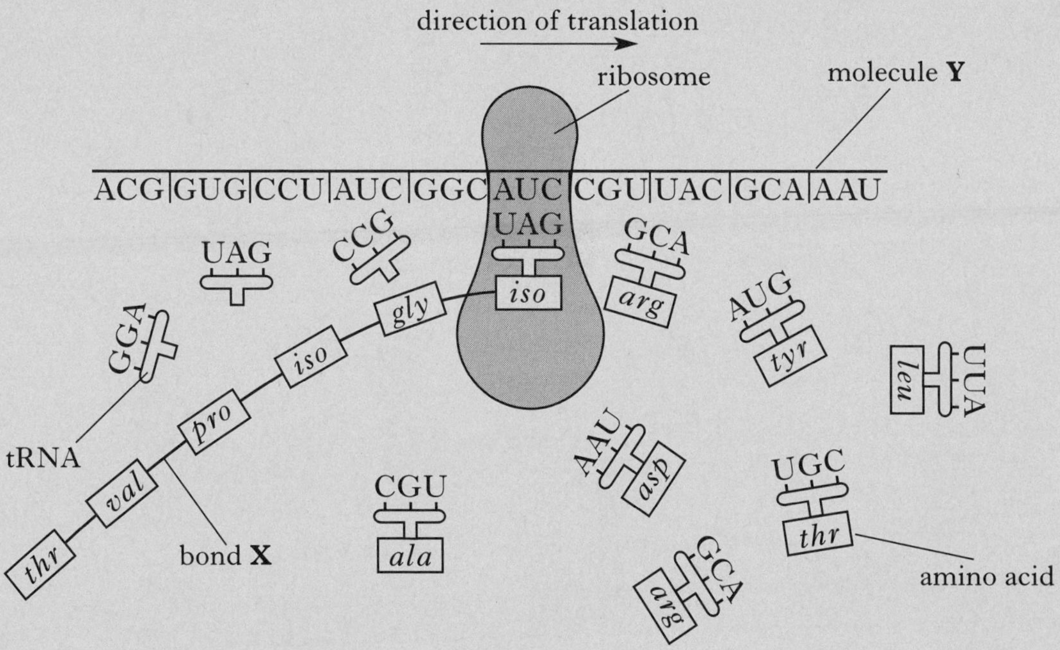

(*a*) Name bond **X** and molecule **Y**.

Bond **X** ______________________ Molecule **Y** ______________________ **2**

(*b*) What term is used to describe the triplet code on the tRNA molecules?

__________________________ **1**

(*c*) Give the abbreviated names of the next **four** amino acids which will be attached to complete the peptide chain.

iso → __________ → __________ → __________ → __________ **1**

(*d*) What sequence of bases on a DNA molecule will code for the amino acid labelled *thr*?

__________________________ **1**

(*e*) Amino acids are added to the peptide chain at the rate of 15 per second.

How long will it take for the complete synthesis of the peptide shown in the diagram above?

__________ s **1**

[Turn over

DO NOT WRITE IN THIS MARGIN

Marks

3. The diagram below shows the inheritance of tongue-rolling ability in a family.

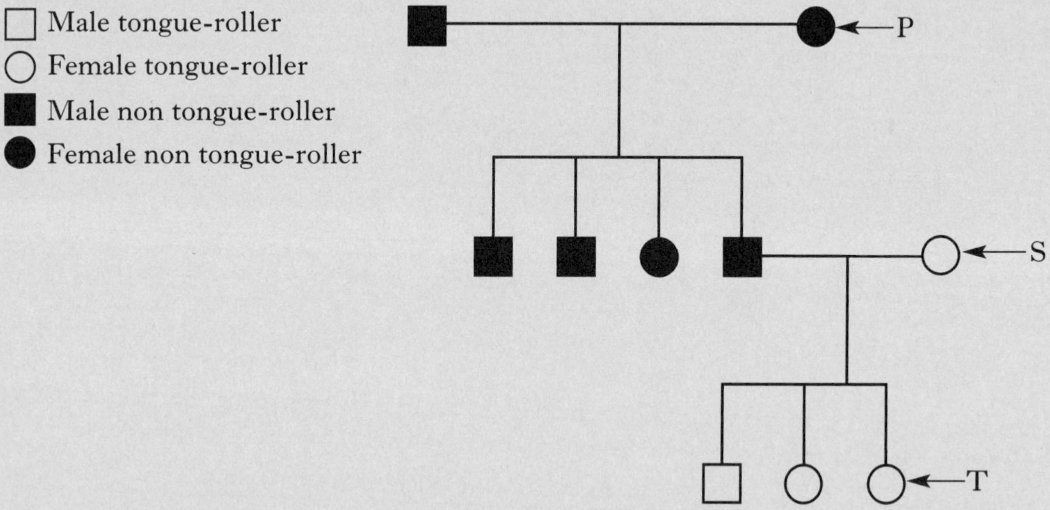

(*a*) (i) Using the symbol **R** to represent the allele for tongue-rolling, and the symbol **r** to represent the allele for non tongue-rolling, state the genotypes of individuals P and T.

P ________ T ________ **1**

(ii) How many individuals, shown in this family tree, have a genotype which is homozygous recessive?

________ **1**

(iii) Place a cross through a symbol in the family tree which represents a heterozygous male. **1**

(iv) Female S is pregnant. Using information from the family tree, is it possible to predict whether the child will be a tongue-roller or not? Give a reason for your answer.

YES/NO ________

Reason ____

____ **1**

(*b*) Some characteristics are controlled by several genes.

(i) State the term used to describe this type of inheritance pattern.

________________________________ **1**

(ii) Which **two** of the following human characteristics show this type of inheritance pattern? *Underline the correct answers.*

blood groups **cystic fibrosis** **height**

haemophilia **skin colour** **1**

DO NOT WRITE IN THIS MARGIN

Marks

4. Some stages in the process of meiosis are shown in the diagrams below.

Only 6 chromosomes are shown in each cell.

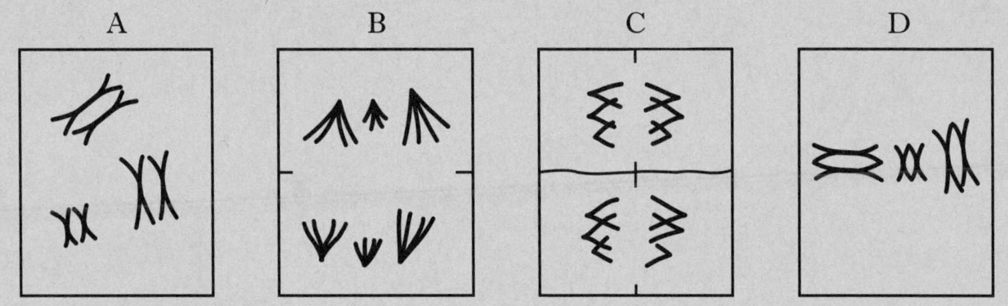

(*a*) (i) Place the stages in the correct order.

________ → ________ → ________ → ________ **1**

(ii) At which stage could chiasmata form?

________ **1**

(iii) Why are chiasmata important?

____

____ **1**

(*b*) The diagram below shows a gamete mother cell and four sperm which would result from meiosis.

(i) Complete the diagram by writing in the normal number of chromosomes found in each of the cells. **1**

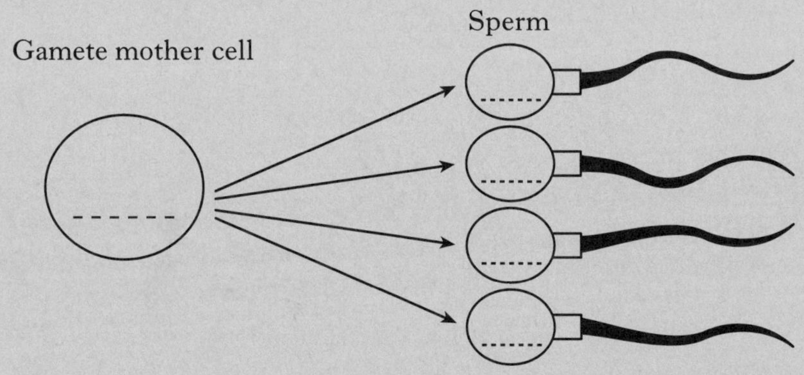

(ii) In how many of these sperm will an **X** chromosome be found?

________ **1**

(iii) Where in the testes does meiosis occur?

________________________ **1**

[Turn over

DO NOT WRITE IN THIS MARGIN

Marks

5. The graph below shows the relative concentrations of three hormones in the plasma of a woman during a normal 28-day menstrual cycle.

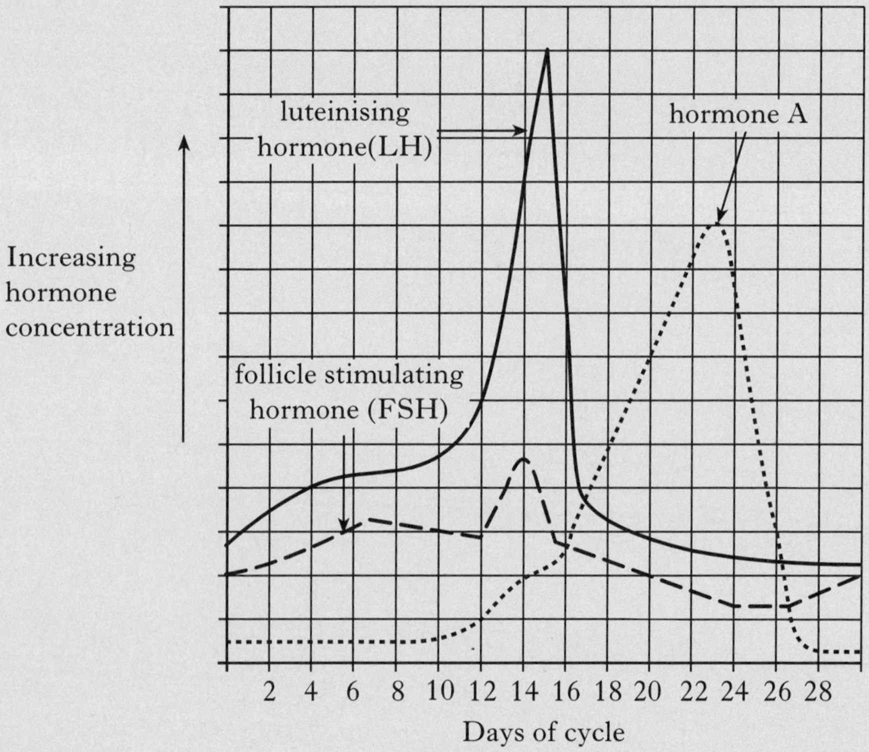

(*a*) Name hormone A.

__________________________ **1**

(*b*) What is the effect of the sudden increase in concentration of luteinising hormone?

__________________________ **1**

(*c*) During which time period is the endometrium likely to reach maximum thickness?

Underline the correct answer.

0–4 days **12–16 days** **22–26 days** **1**

(*d*) In what way would the line showing the concentration of FSH be different if fertilisation took place during this cycle? Give an explanation for your answer.

Difference __________________________

__________________________ **1**

Explanation __________________________

__________________________ **1**

DO NOT WRITE IN THIS MARGIN

Marks

6. The diagram below shows the heart and its associated nerves.

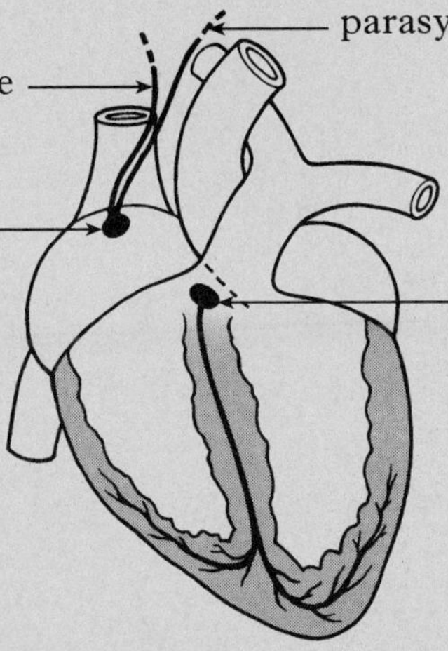

(*a*) On the diagram, mark with an **X** the chamber where the blood pressure is highest during the cardiac cycle. **1**

(*b*) Describe the effect of impulses from the parasympathetic nerve on the heart.

____ **1**

(*c*) (i) Name the part of the heart labelled **B**.

__________________________ **1**

(ii) Describe the role of **B** in the cardiac cycle.

____

____ **1**

(iii) An individual has a heart rate of 75 bpm. How long does one cardiac cycle last?

Space for calculation

____________ s **1**

[Turn over

DO NOT WRITE IN THIS MARGIN

Marks

7. The diagram below shows the structure of a villus.

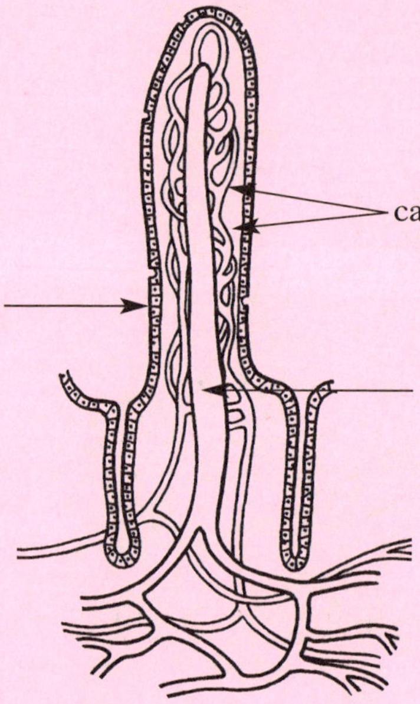

(*a*) (i) Name lymph vessel **X**.

______________________________ **1**

(ii) Describe the role of lymph vessel **X** in the transport of nutrients.

____

____ **1**

(*b*) (i) Vitamin B_{12} is absorbed into the blood capillaries of the villus.

Name the substance which must be present before vitamin B_{12} can be absorbed.

____ **1**

(ii) Which type of body cell requires vitamin B_{12} for its manufacture?

____ **1**

(*c*) Name the blood vessel which transports nutrient-rich blood away from the small intestine.

____ **1**

DO NOT WRITE IN THIS MARGIN

Marks

8. The diagram below shows blood flow through a capillary bed in the pancreas. The area shown is an Islet of Langerhans.

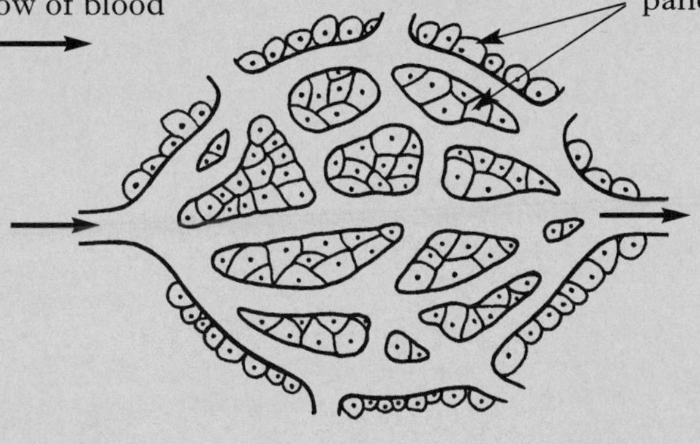

(*a*) After a meal rich in carbohydrate, the composition of the blood changes as it flows through the pancreas.

Complete the table below using the words *increases* or *decreases* to indicate these changes.

Substance	*Change*
Oxygen	
Carbon dioxide	
Insulin	

2

(*b*) (i) What effect does insulin have on the body?

____

____ **1**

(ii) Name a hormone produced by the pancreas which has the opposite effect to insulin.

__________________________ **1**

[Turn over

9. The graphs below contain information about diet in the UK.

Graph 1 shows how UK diet has changed between 1988 and 1998.

Graph 2 shows how Scottish consumption compares with the rest of the UK in 1998.

Graph 1

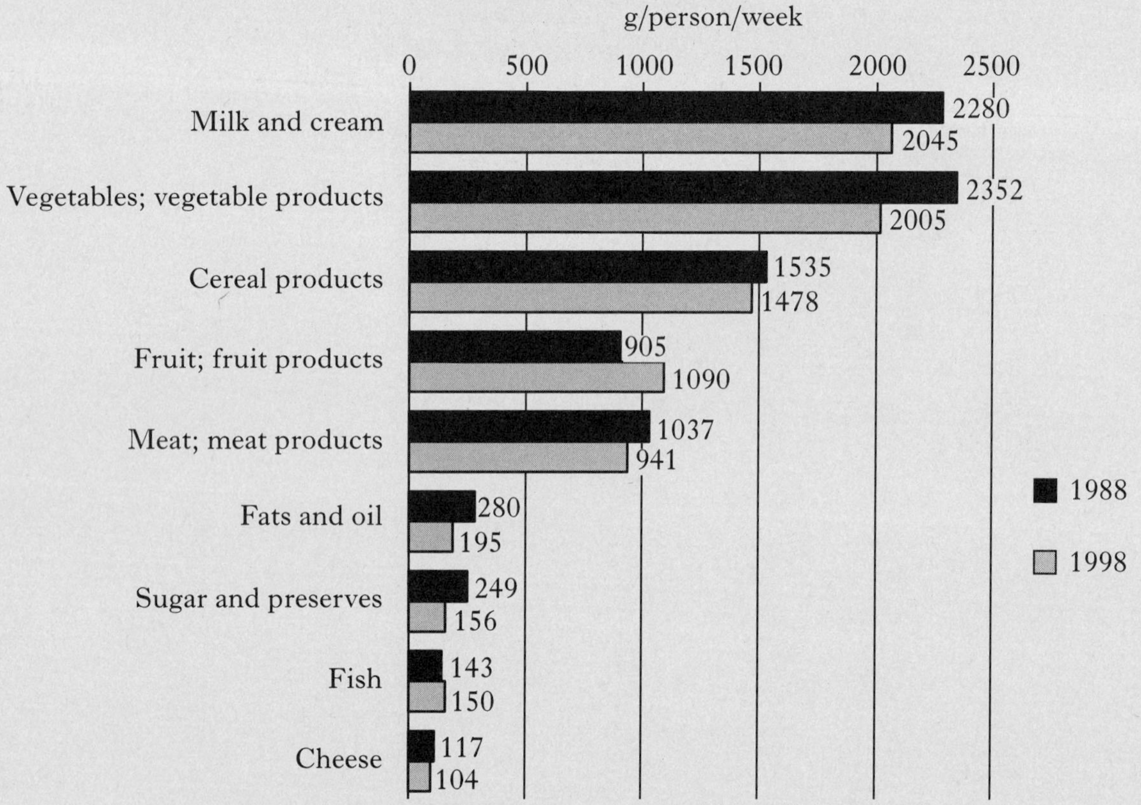

Graph 2

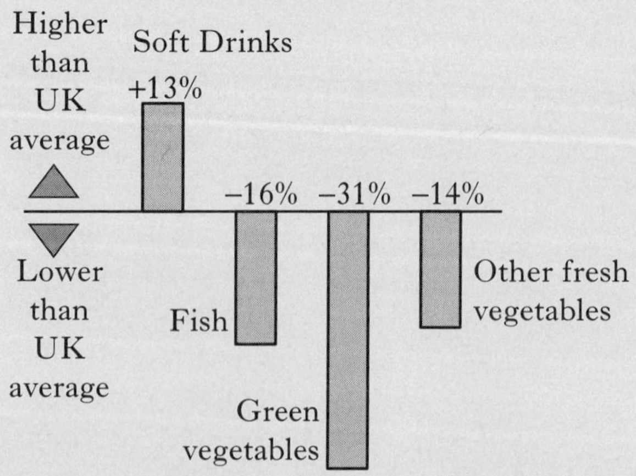

DO NOT WRITE IN THIS MARGIN

9. (continued) *Marks*

(*a*) (i) From **Graph 1**, what general conclusion can be drawn about UK diet in 1988 compared with 1998?

________________________________ **1**

(ii) What additional information would be required in order to draw a general conclusion about the Scottish diet over the same period of time?

________________________________ **1**

(iii) When collecting data to make comparisons of this type, state **two** variables which should be controlled.

1 ________________________________

2 ________________________________ **2**

(*b*) From **Graph 1**, calculate the percentage reduction in milk and cream consumption over the decade.

Space for calculation

________________ % **1**

(*c*) With reference to **both** graphs, calculate the weekly fish consumption in Scotland in 1998.

Space for calculation

________________ g/person/week **1**

(*d*) The incidence of coronary heart disease in the UK over the ten-year period has decreased. From **Graph 1** give **two** pieces of evidence which may contribute to this decrease.

1 ________________________________

2 ________________________________ **1**

(*e*) Information such as this can be used by governments to plan for the future. Suggest what use might be made of the information on diet shown in the graphs.

____

____

____

________________________________ **1**

DO NOT WRITE IN THIS MARGIN

Marks

10. The diagram shows a section through the brain.

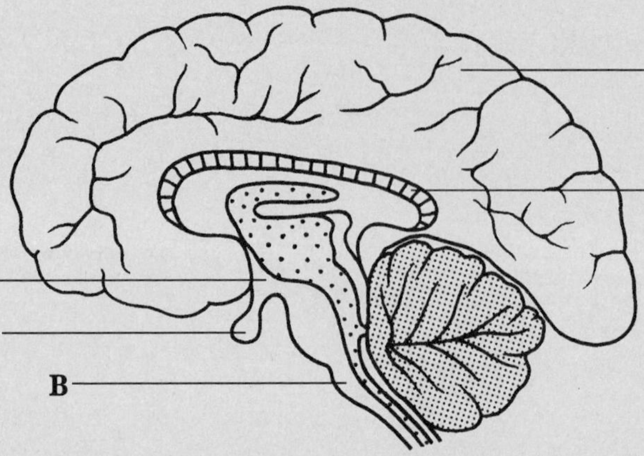

(*a*) Name parts **A** and **B** shown on the diagram.

A ______________________________ **B** ______________________________ 2

(*b*) (i) Name **two** areas of the cerebrum in which functions are localised.

______________________________ and ______________________________ 1

(ii) Explain how the convoluted surface of the cerebrum contributes to its function.

____

____ 1

(iii) What is the function of the corpus callosum?

____

____ 1

(*c*) Parts of the brain are involved in memory storage. Complete the following sentences which relate to memory loss, using words from the list below.

Alzheimer's	**noradrenaline**	**limbic**
Huntingdon's	**acetylcholine**	**lymphatic**

A disorder particularly associated with memory loss is ____________________ disease. This disorder is due to the disappearance of cells which produce the neurotransmitter ____________________ in the ____________________ system of the brain. 2

DO NOT WRITE IN THIS MARGIN

Marks

10. (continued)

(*d*) The hormone ADH is produced by the pituitary gland.

Describe the role of ADH in restoring water balance after excess water has been lost from the body.

____

____

____

____

____ **3**

[Turn over

DO NOT WRITE IN THIS MARGIN

Marks

11. A student carried out an experiment to investigate how quickly three groups of pupils completed a finger maze.

The table below shows the fastest times obtained by each group of pupils.

	Fastest time achieved in trials (s)		
	8 year-olds	*12 year-olds*	*16 year-olds*
	11	7	7
	12	8	8
	9	6	7
	6	10	9
	11	8	6
	10	7	–
	12	12	–
	13	6	–
	–	7	–
	–	6	–
average	**10·5**		**7·4**

(*a*) Calculate the average time for the 12 year-old pupils and write your answer in the table above. **1**

(*b*) Construct a **bar graph** to show the **average** times for each age group of pupils.

(Additional graph paper, if required, will be found on page 30.)

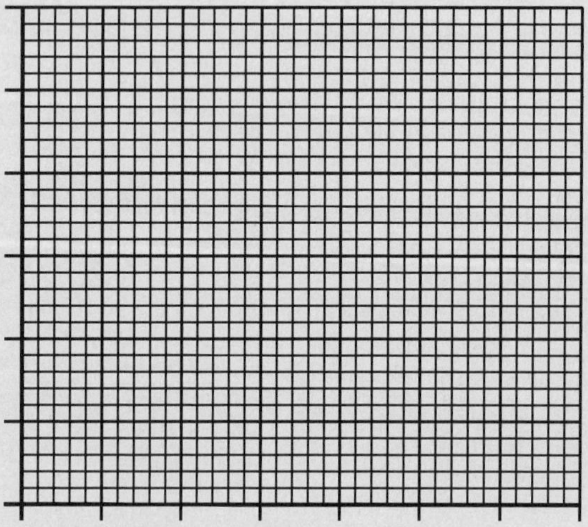

2

DO NOT WRITE IN THIS MARGIN

11. (continued) *Marks*

(*c*) (i) Explain why these average results are not reliable.

____

____ **1**

(ii) Identify another factor which would have to be considered in the design of this experiment to ensure that valid conclusions can be drawn.

____

____ **1**

(*d*) It was found that some of the pupils performed better when they were being watched by other pupils. What term is used to describe this effect?

________________________ **1**

[Turn over

DO NOT WRITE IN THIS MARGIN

Marks

12. The graph below shows two survival curves for the UK population.

The curves plot the number of people at each age who are still alive.

One curve is from the year 1800 and the other from the year 1960.

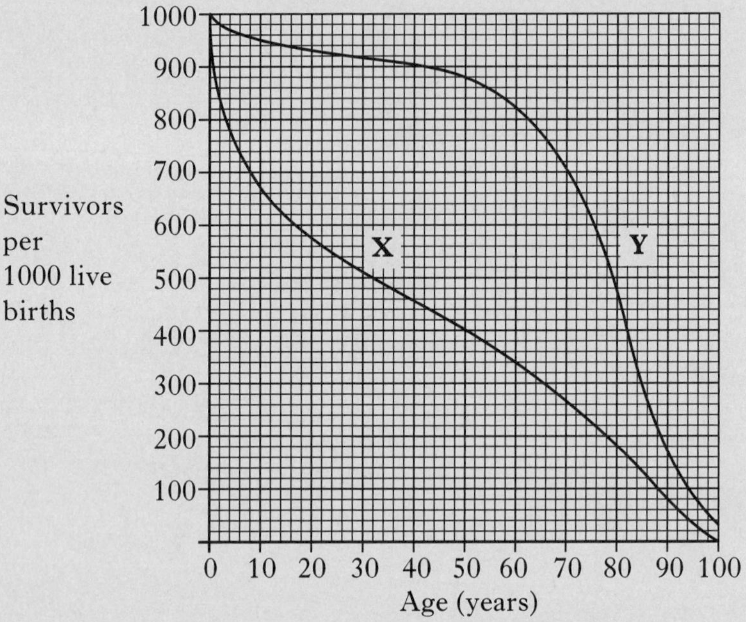

(*a*) Complete the table to identify each curve and give a reason for your selection.

Curve	*Year*
X	
Y	

Reason for selection

____

______________________________ **1**

(*b*) Suggest **two** reasons for the difference in the curves between the ages of 0 and 10.

1 ______________________________

____

2 ______________________________

______________________________ **1**

DO NOT WRITE IN THIS MARGIN

Marks

12. (continued)

(*c*) For each curve, express the numbers still alive at the age of 50 as a percentage of their original population.

Space for calculation

X ________________ % **Y** ________________ % 1

(*d*) For each of the curves identify the ten-year period during which there is the highest mortality rate.

X ________________ years **Y** ________________ years 1

[Turn over

13. The graphs below relate to fertiliser use by farmers, and nitrate pollution of two rivers from 1930 to 1990.

Graph 1. The use of nitrogen (N) and phosphorus (P) in fertiliser.

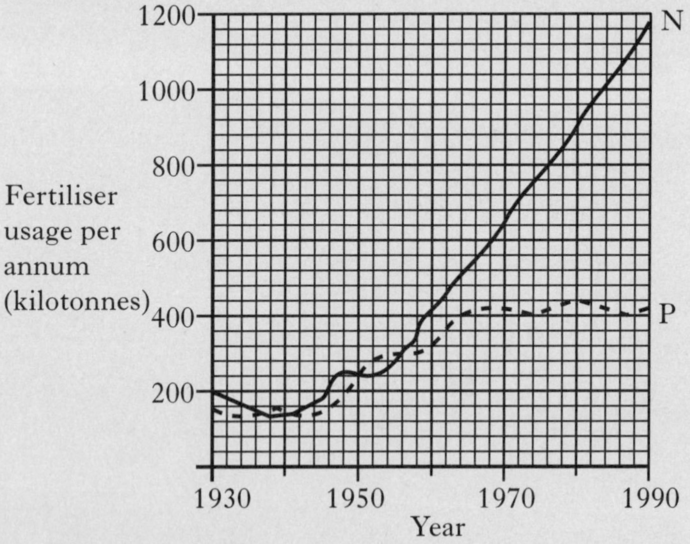

Graph 2. The nitrate content of the rivers Dee and Tay.

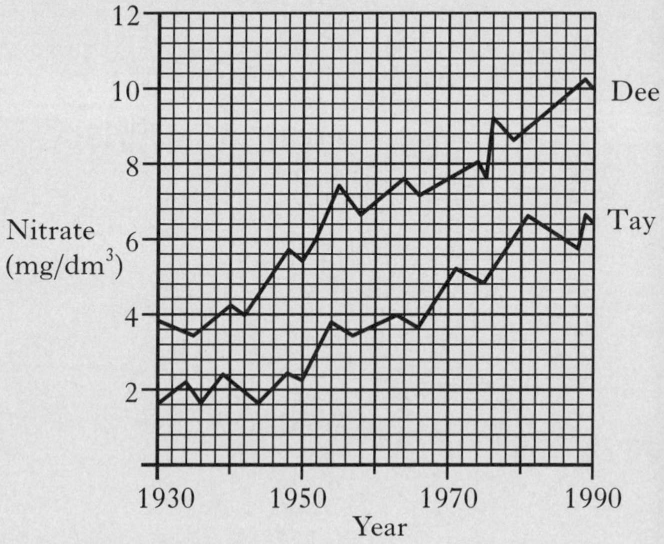

N – Nitrogen
P – Phosphorus

DO NOT WRITE IN THIS MARGIN

Marks

(*a*) From **Graph 1**, compare the change in use of nitrogen and phosphorus in fertiliser from 1930 to 1990.

____

____

____

____ **2**

(*b*) From **Graph 2**, what was the nitrate concentration in the river Dee in the year 1970?

______________ **1**

(*c*) What evidence is there, from comparison of both graphs, that agricultural use of fertiliser is linked to nitrate pollution of the Dee and Tay?

____

____ **1**

DO NOT WRITE IN THIS MARGIN

13. (continued) *Marks*

(*d*) Describe **two** effects of increased nitrate pollution of fresh water.

1 ____

____

2 ____

____ **2**

(*e*) Name a source of nitrate and phosphate pollution other than farmland.

________________________ **1**

(*f*) Describe **one** way in which phosphorus is important to the structure or function of a cell.

____

____ **1**

[Turn over

DO NOT WRITE IN THIS MARGIN

Marks

SECTION C

Both questions in this section should be attempted.

Note that each question contains a choice.

Questions 1 and 2 should be attempted on the blank pages which follow.

Supplementary sheets, if required, may be obtained from the invigilator.

Labelled diagrams may be used where appropriate.

1. Answer **either** A **or** B.

A. Give an account of DNA under the following headings:

(i) DNA structure; **7**

(ii) DNA replication. **3**

(10)

OR

B. Give an account of immunity under the following headings:

(i) B-lymphocytes and T-lymphocytes; **7**

(ii) Macrophages. **3**

(10)

In question 2 ONE mark is available for coherence and ONE mark is available for relevance.

2. Answer **either** A **or** B.

A. Give an account of the life history of a red blood cell. **(10)**

OR

B. Give an account of the influence of hormones on the growth and development of boys. **(10)**

[*END OF QUESTION PAPER*]

DO NOT WRITE IN THIS MARGIN

SPACE FOR ANSWERS

DO NOT WRITE IN THIS MARGIN

SPACE FOR ANSWERS

ADDITIONAL GRAPH PAPER FOR QUESTION 11(*b*)

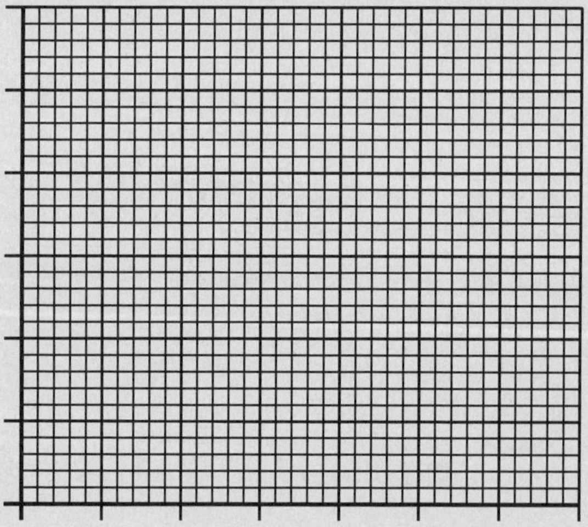

2003 HIGHER

FOR OFFICIAL USE

Total for Sections B & C

X009/301

NATIONAL QUALIFICATIONS 2003

MONDAY, 26 MAY
1.00 PM – 3.30 PM

HUMAN BIOLOGY HIGHER

Fill in these boxes and read what is printed below.

Full name of centre

Town

Forename(s)

Surname

Date of birth
Day Month Year

Scottish candidate number

Number of seat

SECTION A—Questions 1–30

Instructions for completion of Section A are given on page two.

SECTIONS B AND C

1 (a) All questions should be attempted.

(b) It should be noted that in **Section C** questions 1 and 2 each contain a choice.

2 The questions may be answered in any order but all answers are to be written in the spaces provided in this answer book, and must be written clearly and legibly in ink.

3 Additional space for answers and rough work will be found at the end of the book. If further space is required, supplementary sheets may be obtained from the invigilator and should be inserted inside the **front** cover of this book.

4 The numbers of questions must be clearly inserted with any answers written in the additional space.

5 Rough work, if any should be necessary, should be written in this book and then scored through when the fair copy has been written.

6 Before leaving the examination room you must give this book to the invigilator. If you do not, you may lose all the marks for this paper.

SCOTTISH QUALIFICATIONS AUTHORITY

SECTION A

Read carefully

1 Check that the answer sheet provided is for Human Biology Higher (Section A).

2 Fill in the details required on the answer sheet.

3 In this section a question is answered by indicating the choice A, B, C or D by a stroke made in **ink** in the appropriate place in the answer sheet—see the sample question below.

4 For each question there is only **one** correct answer.

5 Rough working, if required, should be done only on this question paper—or on the rough working sheet provided—**not** on the answer sheet.

6 At the end of the examination the answer sheet for Section A **must** be placed **inside** this answer book.

Sample Question

The digestive enzyme pepsin is most active in the

A mouth

B stomach

C duodenum

D pancreas.

The correct answer is **B**—stomach. A **heavy** vertical line should be drawn joining the two dots in the appropriate box in the column headed **B** as shown in the example on the answer sheet.

If, after you have recorded your answer, you decide that you have made an error and wish to make a change, you should cancel the original answer and put a vertical stroke in the box you now consider to be correct. Thus, if you want to change an answer D to an answer B, your answer sheet would look like this:

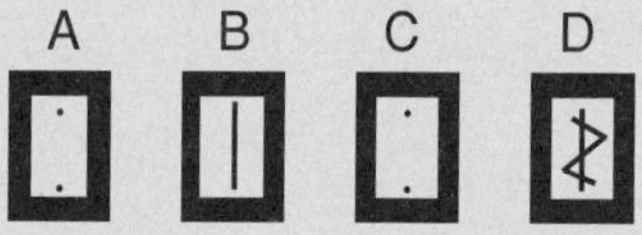

If you want to change back to an answer which has already been scored out, you should enter a tick (✓) to the **right** of the box of your choice, thus:

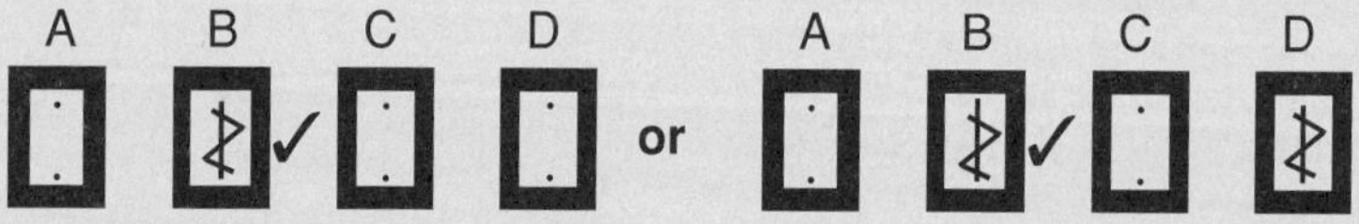

SECTION A

All questions in this section should be attempted.

Answers should be given on the separate answer sheet provided.

1. The diagram represents part of a molecule of DNA on which a molecule of RNA is being synthesised.

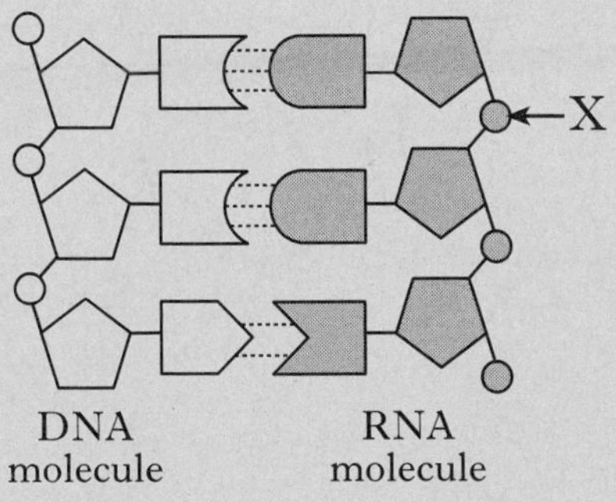

What does component X represent?

A Ribose sugar

B Deoxyribose sugar

C Phosphate

D Ribose phosphate

2. DNA controls the activities of a cell by coding for the production of

A proteins

B carbohydrates

C amino acids

D bases.

3. The diagram shows a stage in the synthesis of part of a polypeptide.

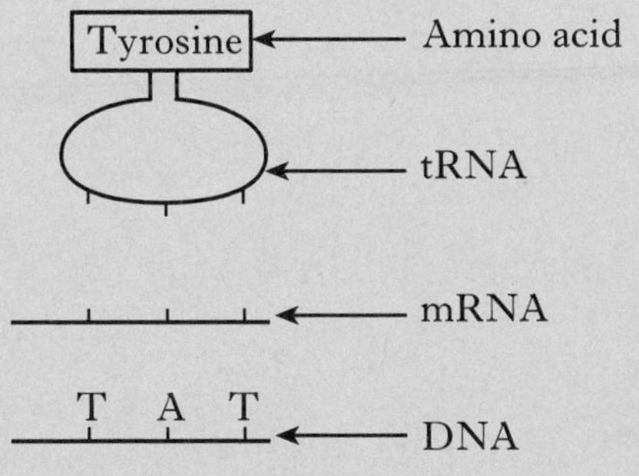

Identify the triplet codes for the amino acid tyrosine.

	mRNA	*tRNA*
A	ATA	UAU
B	UAU	AUA
C	AUA	UAU
D	ATA	TAT

4. The base sequence of a short piece of DNA is shown below.

A G C T T A C G

During replication, an inversion mutation occurred on the complementary strand synthesised on this piece of DNA.

Which of the following is the mutated complementary strand?

A T C G A A T G A

B A G C T T A G C

C T C G A A T C G

D T C G A A T G C

5. The graph shows the effect of substrate concentration on the rate of an enzyme-controlled reaction.

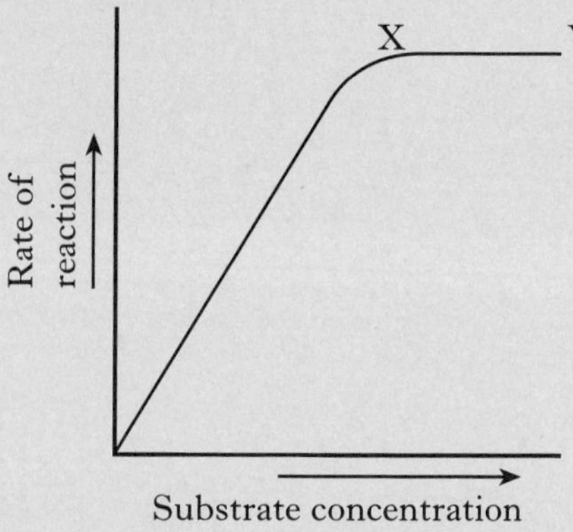

The graph levels out between points X and Y because the

A enzyme is denatured

B active sites are saturated with substrate

C enzyme is inhibited

D enzyme is activated.

6. When a protease enzyme is added to an amylase solution, which of the following could be produced?

A Amino acids

B Maltose

C Glucose

D Glycerol

7. The diagram below represents stages in tissue respiration.

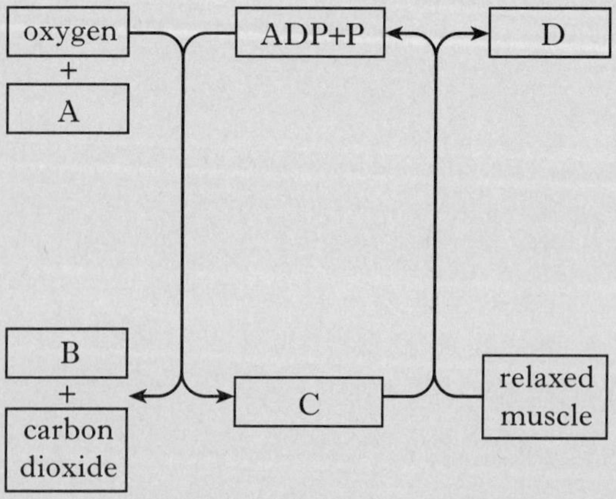

Which box represents ATP?

8. A piece of muscle was cut into three strips X, Y and Z and treated as described in the table.

Their final lengths were then measured.

Muscle strip	*Solution added to muscle*	*Muscle length* (mm)	
		Start	*After 10 minutes*
X	1% glucose	50	50
Y	1% ATP	50	45
Z	1% ATP boiled and cooled	50	46

From the data it may be deduced that

A ATP is not an enzyme

B muscles contain many mitochondria

C muscles synthesise ATP in the absence of glucose

D muscles do not use glucose as a source of energy.

9. Which line in the table has pairs of statements which are true with regard to aerobic respiration and anaerobic respiration in human muscle tissue?

	Aerobic respiration	*Anaerobic respiration*
A	There is a net gain of ATP	Carbon dioxide is not produced
B	There is a net gain of ATP	Oxygen is used up
C	Carbon dioxide is evolved	There is a net loss of ATP
D	Lactic acid is formed	Ethanol is formed

10. The graph below shows changes which occur in the masses of protein, fat and carbohydrate in a boy's body during five weeks of starvation.

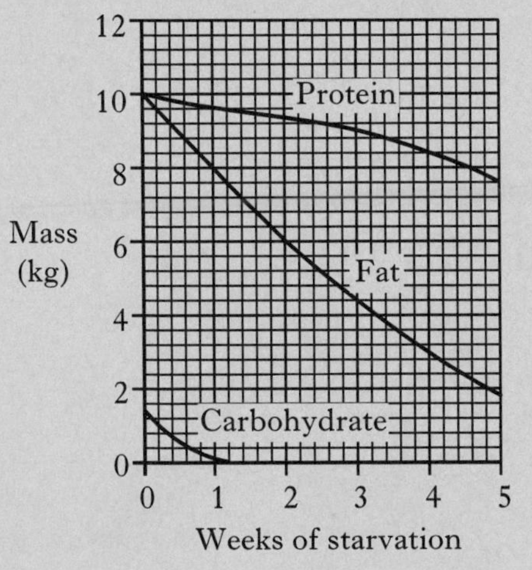

The boy weighs 60 kg. Predict his weight after two weeks without food.

A 43 kg

B 50 kg

C 54 kg

D 57 kg

11. Which of the following will result in the gain of active immunity by the body?

A Transfer of antibodies across the placenta

B Injection of antitoxin

C Suckling of breast milk

D Invasion by viruses

12. Which of the following reactions describes autoimmunity?

A The production of antibodies in response to infection

B The rapid production of antibodies in response to reinfection

C The production of antibodies in response to immunisation

D The production of antibodies in response to the body's own cells

13. Two parents, one of blood group A and the other of blood group B, have four children.

The phenotypes of the children are all different – blood group A, blood group B, blood group AB and blood group O.

What are the genotypes of the parents?

A $AA \times BB$

B $AO \times BO$

C $AA \times BO$

D $AO \times BB$

14. Which one of the following statements about sex-linked traits is true?

A A female transmits her sex-linked traits to her daughters only.

B A male transmits his sex-linked traits to his sons only.

C A male transmits his sex-linked traits to his grandchildren via his daughters only.

D A female transmits her sex-linked traits to her grandchildren via her sons only.

15. Colour blindness is a sex-linked trait. A man with normal vision marries a woman with normal vision. They have a son who is colour blind.

What is the chance of their next son being colour blind?

A no chance

B 1 in 2

C 1 in 3

D 1 in 4

16. The function of the seminal vesicles is to

A produce nutrients for sperm

B allow sperm to mature

C store sperm temporarily

D produce testosterone.

[Turn over

17. Which of the following is the sequence of events following fertilisation?

A Cleavage → Differentiation → Implantation

B Implantation → Differentiation → Cleavage

C Differentiation → Implantation → Cleavage

D Cleavage → Implantation → Differentiation

18. The electrocardiogram shown below records the beat of a human heart.

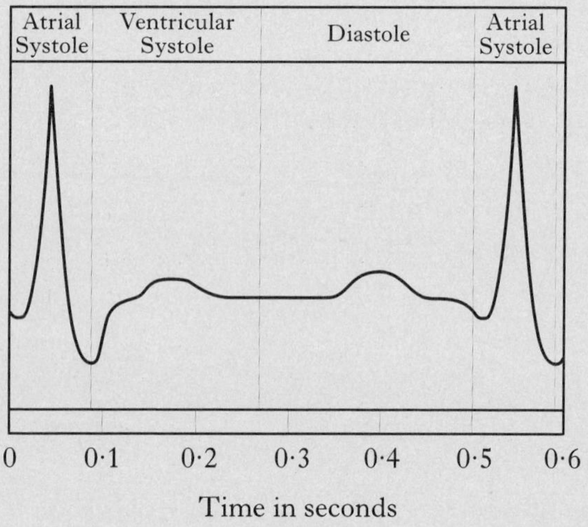

What is the heart rate?

A 60 beats/minute

B 70 beats/minute

C 75 beats/minute

D 120 beats/minute

19. Which of the following is **not** a function of the lymphatic system?

A It returns excess tissue fluid to the blood.

B It transports fat from the small intestine.

C It destroys bacteria.

D It causes the clotting of blood at wounds.

20. The following data refer to the breathing of an athlete resting and just after a race.

	Breathing Rate (per minute)	*Tidal Volume* (litres)	*% Carbon dioxide in exhaled air*
Resting	12	0·3	5
After race	24	1	5

Assuming the rate of breathing remains constant, what would be the volume of carbon dioxide breathed out during the first two minutes after exercise?

A 0·18 litres

B 0·36 litres

C 1·2 litres

D 2·4 litres

21. Which line in the table below identifies correctly the sites of secretion of the hormones ADH and glucagon?

	ADH	*Glucagon*
A	pituitary gland	liver
B	kidney	liver
C	kidney	pancreas
D	pituitary gland	pancreas

22. The graph below shows the blood glucose concentrations of two women before and after each swallowed 50 g of glucose.

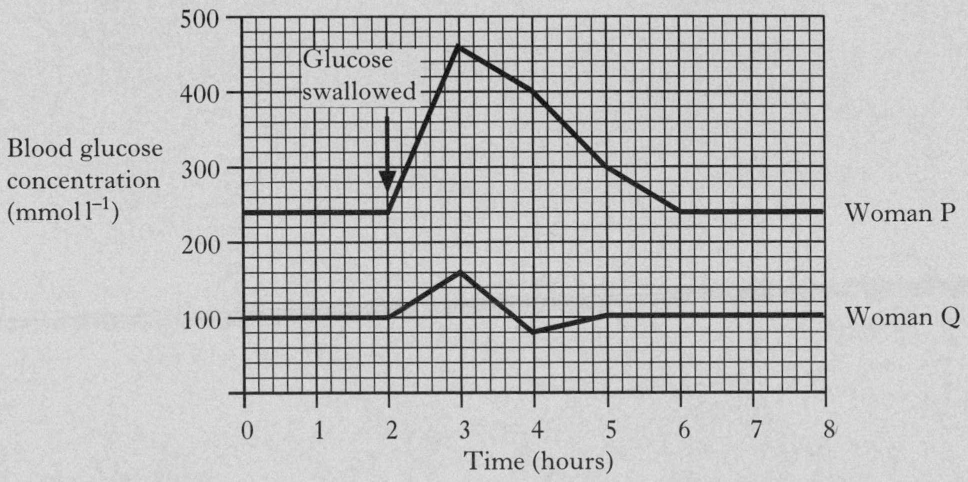

When did the rate of change of blood glucose concentration of the two women differ most?

A Between hours 2 and 3

B Between hours 3 and 4

C Between hours 4 and 5

D Between hours 5 and 6

23. Both nerves and hormones are involved in the control of homeostatic mechanisms.

In which of the following are the homeostatic mechanisms related correctly to their principal methods of control?

	Osmoregulation	*Body Temperature*
A	Nerves	Hormones
B	Hormones	Nerves
C	Nerves	Nerves
D	Hormones	Hormones

24. The human cerebrum has a highly convoluted surface. This increased surface area allows an

A increase in the types of neurones present

B increased blood supply to the brain

C increased number of interconnections between neurones

D increase in the amount of white matter on the surface.

25. Which parts of the body are controlled by the largest motor area of the cerebrum?

A Hands and lips

B Feet and hands

C Legs and feet

D Legs and arms

26. Vision in dim light is improved by the rods having

A diverging neural pathways

B converging neural pathways

C reflex neural pathways

D peripheral neural pathways.

27. The retrieval of information from long term memory is often aided by remembering the situation in which the information was encoded. This is described as using

A contextual cues

B chunking techniques

C rehearsal methods

D memory span.

28. The transformation of information into a form that memory can accept is called

A shaping

B retrieval

C encoding

D storage.

[Turn over

29. The graph shows how the algal population in a freshwater lake is affected by changes in the nitrate concentration in the water and by the changes in light intensity over the year.

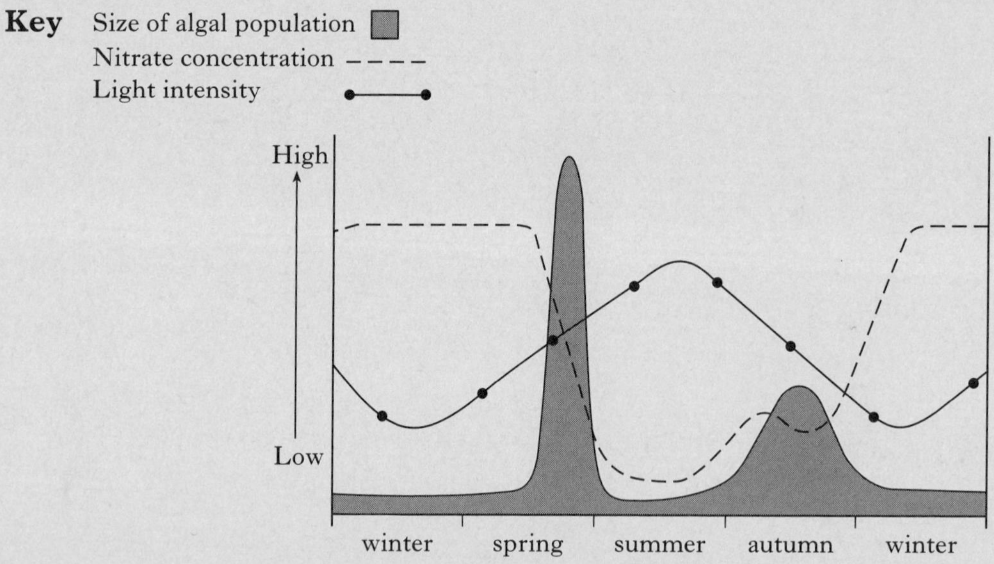

Which of the following statements may be deduced from the graph?

A The increase in the algal population in the spring is triggered by a high concentration of nitrate.

B The changes in size of the algal population are related directly to changes in light intensity.

C The increasing concentration of nitrate in the autumn is followed by an increase in the algal population.

D The size of the algal population is inversely proportional to light intensity.

30. The diagram below shows the population of a country as a percentage distribution by age.

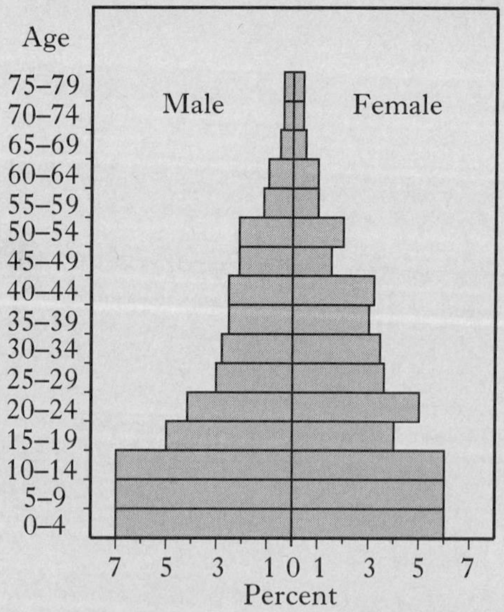

What percentage of the population is under 15 years of age?

A 21%

B 39%

C 48%

D 49%

Candidates are reminded that the answer sheet MUST be returned INSIDE this answer booklet.

[Turn over for Section B]

DO NOT WRITE IN THIS MARGIN

Marks

SECTION B

All questions in this section should be attempted.

1. The diagram shows the sequence of events as a white blood cell engulfs and destroys a bacterium.

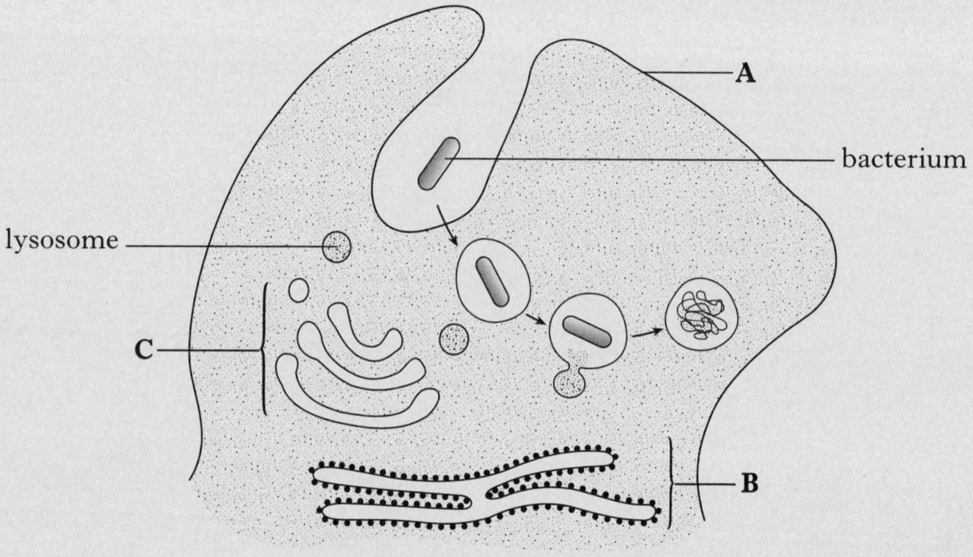

(*a*) Name the structures **A**, **B** and **C**.

A ______________________

B ______________________

C ______________________ **2**

(*b*) (i) What is the name given to this process of engulfing the bacterium?

______________________ **1**

(ii) Describe the roles of structures **B** and **C** in the cell.

B ______________________

______________________ **1**

C ______________________

______________________ **1**

(*c*) What does the lysosome contain?

______________________ **1**

DO NOT WRITE IN THIS MARGIN

Marks

1. (continued)

(*d*) (i) Analysis of a blood sample yielded the following blood cell counts.

Cell type	*Number per mm³* (× 1000)
White	8
Red	5600

Express as a simple ratio the number of white cells to red cells in this blood sample.

Space for calculation

______ : ______ **1**

(ii) Predict how the proportion of white cells to red cells would change if a person was suffering from influenza.

______________________________ **1**

(*e*) Explain how the shape of the red blood cell is related to its function.

____

____

____

______________________________ **2**

[Turn over

DO NOT WRITE IN THIS MARGIN

Marks

2. The graph below shows oxygen-haemoglobin dissociation curves at 37 °C and at 38 °C.

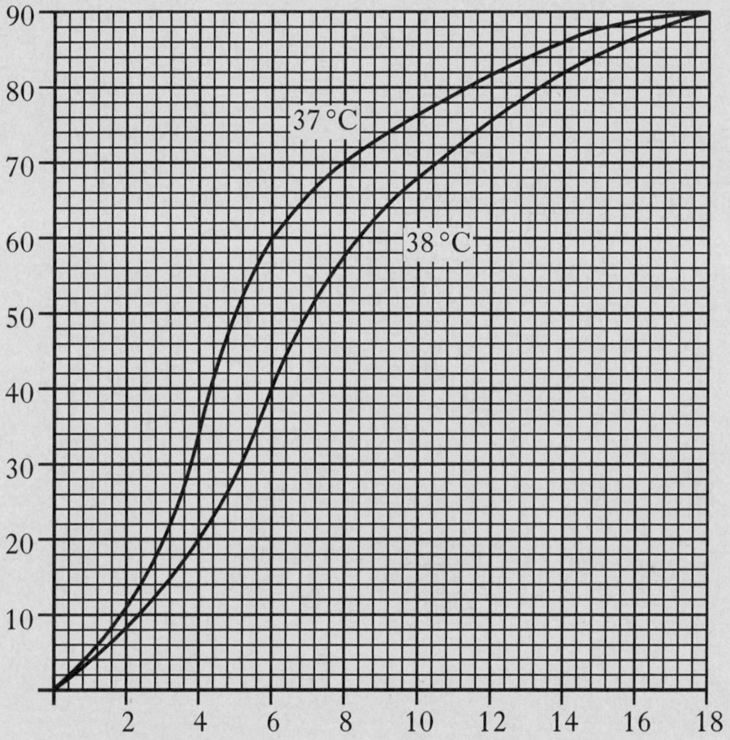

(*a*) (i) Complete the table below to show the change in percentage oxygen saturation of haemoglobin at 37 °C and 38 °C when the partial pressure drops from 18 to 6 kPa.

Partial pressure kPa	*Percentage oxygen saturation of haemoglobin*	
	37 °C	38 °C
18		
6		
change		

2

(ii) Explain why this change in percentage oxygen saturation of haemoglobin improves the efficiency of working muscles.

____

____

______________________________ 2

(*b*) The partial pressure of oxygen in fresh air is 20 kPa. The partial pressure of oxygen in the alveoli is 16 kPa. Explain why there is a lower value for oxygen in the alveoli.

____

______________________________ 1

DO NOT WRITE IN THIS MARGIN

Marks

3. Thalassaemia is an inherited blood disorder in which haemoglobin is affected. The condition illustrates incomplete dominance in which the recessive allele has a partial effect. Heterozygous individuals show mild symptoms.

The diagram below shows the incidence of thalassaemia in three generations of a family.

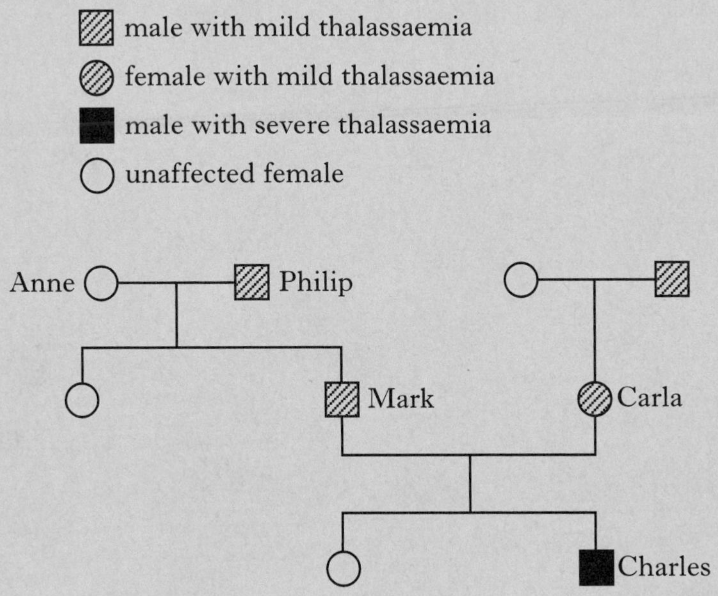

(*a*) Using the symbol Hb^A to represent the allele for normal haemoglobin, and the symbol Hb^B to represent the recessive allele, complete the table to show the genotypes of Anne, Philip and Charles.

Individual	*Genotype*
Anne	
Philip	
Charles	

2

(*b*) Mark and Carla have a third child. What is the percentage chance that the child will have the same genotype as the parents?

Space for calculation

__________ % 1

(*c*) Haemoglobin is found in red blood cells. Where in the body are red cells manufactured and destroyed?

Manufactured ____________________ Destroyed ____________________ 2

[Turn over

DO NOT WRITE IN THIS MARGIN

Marks

4. The diagram shows part of an enzyme molecule.

P represents a molecule in the chain which forms the primary structure of the protein.

Q is a bond which links these molecules.

R is a bond which maintains the secondary structure of the protein.

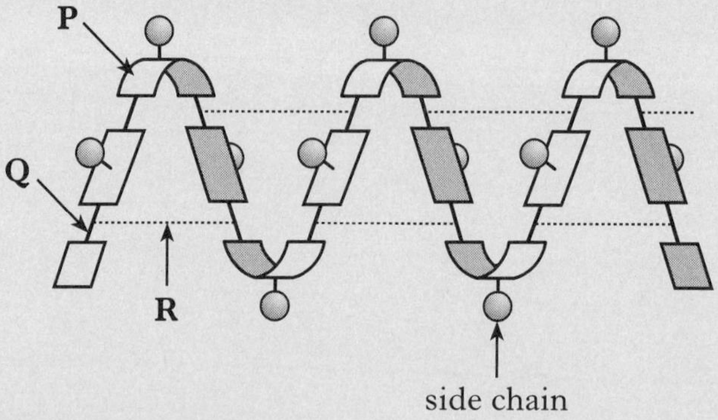

(*a*) (i) Identify molecule **P**.

________________________ 1

(ii) Name bonds **Q** and **R**.

Q __________________ **R** __________________ 2

(*b*) (i) Why are some digestive enzymes produced in an inactive form?

____

____________________________________ 1

(ii) Give an example of a substance which can act as an enzyme activator.

____________________________________ 1

(*c*) The table below contains information about the effects of environmental factors on enzymes. Complete the table.

Factor	*Type of change*	*Effect on enzyme structure*
high temperature		alters active site
	mutation	

3

DO NOT WRITE IN THIS MARGIN

Marks

5. The diagram below represents two stages in the chemistry of respiration in a cell which is respiring aerobically.

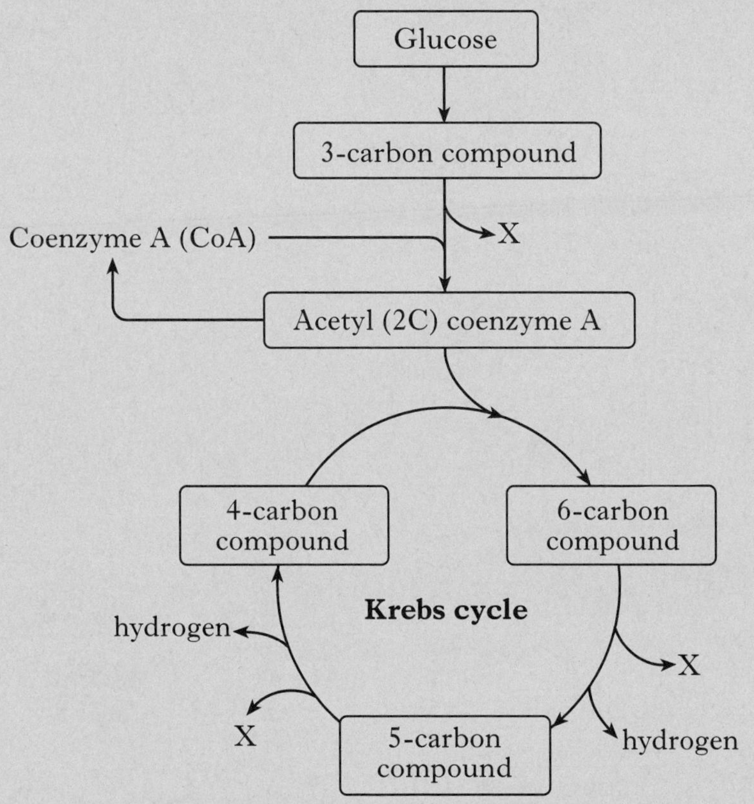

(*a*) Name the 3-carbon compound and the 6-carbon compound.

3C ______________________________

6C ______________________________ **2**

(*b*) Draw a line across the diagram to indicate where this series of reactions would stop if oxygen were not available. **1**

(*c*) State the precise location of the Krebs cycle within the cell.

______________________________ **1**

(*d*) Complete the table below to name product **X** and to describe what happens to each of the products.

Product	*Fate of product*
X ______________	
Hydrogen	

2

DO NOT WRITE IN THIS MARGIN

Marks

6. The diagram below shows the liver with its blood supply and an associated organ.

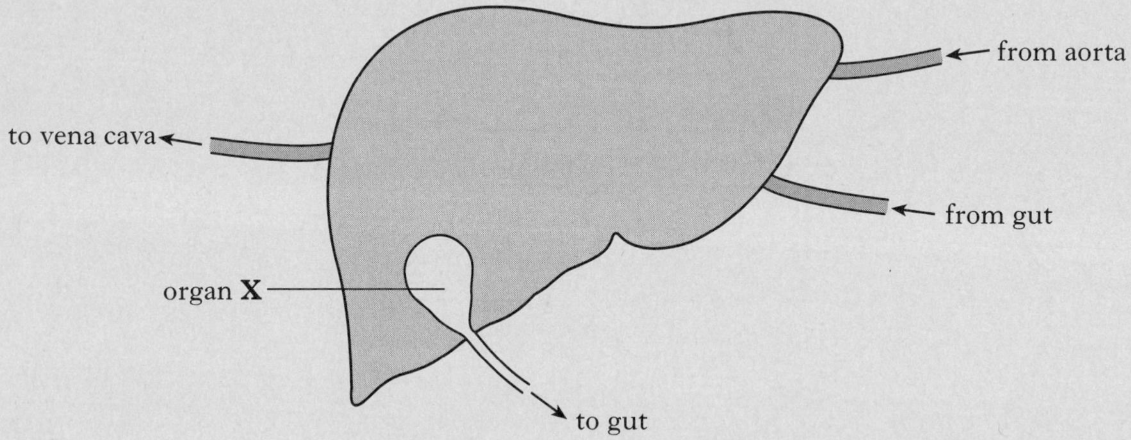

(*a*) Name the liquid stored in organ **X**.

______________________________ 1

(*b*) Complete the table to identify the blood vessels carrying blood to and from the liver and the type of blood carried by each vessel.

Blood supply	*Name of blood vessel*	*Deoxygenated or oxygenated blood*
from aorta		
from gut		
to vena cava		

3

(*c*) Complete the following sentences by <u>underlining</u> one option in each set of brackets.

Deamination occurs in the {gall bladder / kidney / liver}. During this process excess {nucleic acids / amino acids / fatty acids} are broken down. A waste product of this process is {urea / glycogen / glucose} which is carried in the blood to the {urinary bladder / kidney / liver} where it is removed from the blood. 2

(*d*) Name a hormone involved in the processing of carbohydrate in the liver.

______________________________ 1

DO NOT WRITE IN THIS MARGIN

Marks

7. An investigation was carried out to determine the urea content of three samples taken from a healthy individual. The three fluids sampled were:

A plasma from the renal artery

B plasma from the renal vein

C urine from the urethra.

One urease tablet was added to each sample as shown in the diagram below.

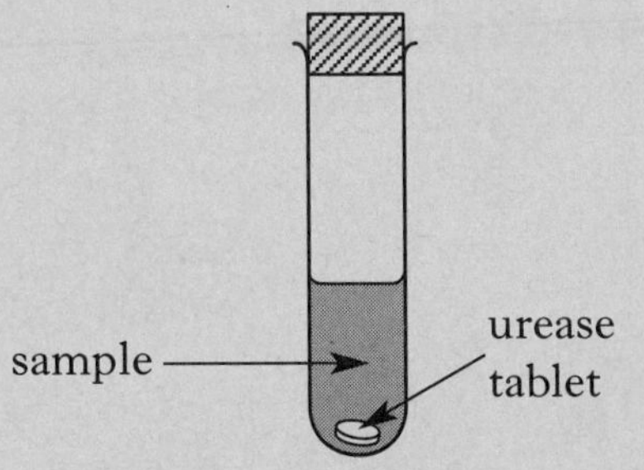

The urease catalyses the breakdown of urea to ammonium carbonate.

urea + water $\xrightarrow{\textit{urease}}$ **ammonium carbonate**

Once the reaction had finished, the samples were analysed to determine the concentration of ammonium carbonate.

(*a*) List **two** variables which would have to be kept constant for a valid comparison of the three samples.

1 ______________________________

2 ______________________________ **2**

(*b*) The table below shows the results of this investigation.

Complete the table to identify the three samples.

Fluid sample (A, B or C)	*Ammonium carbonate concentration* (g/litre)
	0·16
	16·7
	0·52

1

(*c*) Suggest how the investigation could be improved to ensure the reliability of the results.

____

______________________________ **1**

[Turn over

Marks

8. The diagram below shows stages in the control of a person's body temperature.

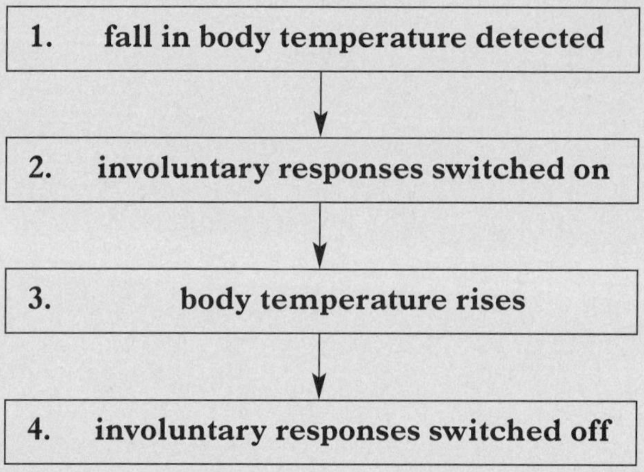

(*a*) Where in the body is the temperature monitoring centre located?

______________________________ **1**

(*b*) State **two** involuntary responses which may be switched on in this individual.

1 ______________________________

2 ______________________________ **2**

(*c*) Explain the role of negative feedback control in this process.

____

____

______________________________ **2**

(*d*) The table below shows the surface area and mass of a baby and an adult.

	Surface area (m^2)	*Mass* (kg)
Baby	0·2	4
Adult	2·0	80

With reference to the table, explain why babies are more susceptible to hypothermia than adults.

____

____

____

______________________________ **2**

DO NOT WRITE IN THIS MARGIN

Marks

9. The graph shows increases in brain volume at four stages of human evolution over the last four million years.

The bars indicate the range of volumes and the mid (median) volume.

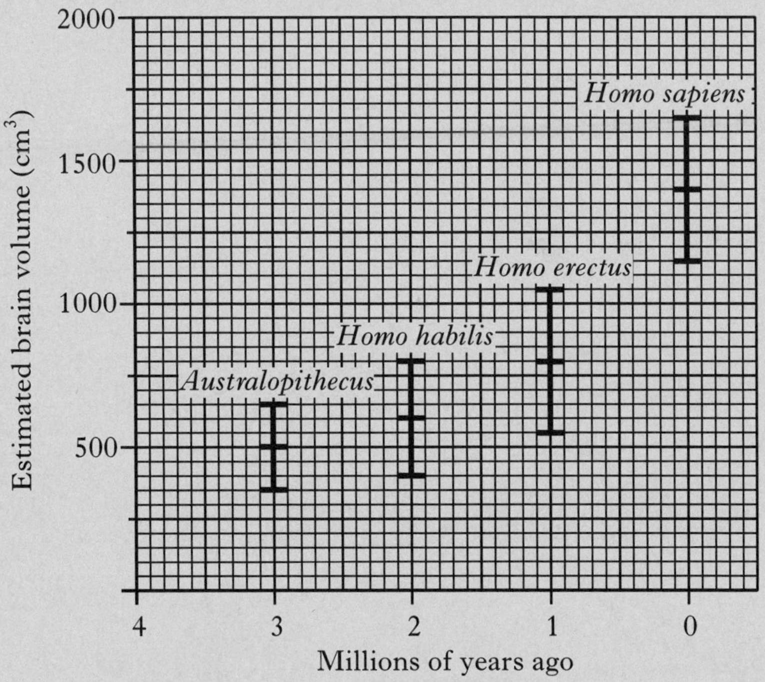

(*a*) State the range of brain volume for *Homo habilis*.

_______________ to _______________ cm^3 **1**

(*b*) Complete the table below for *Homo sapiens*.

Species	*Median volume* (cm^3)	*Percentage increase*
Australopithecus	500	–
Homo habilis	600	20%
Homo erectus	800	33%
Homo sapiens		

2

(*c*) Name the part of the brain which

(i) contributes most to brain volume;

______________________________ **1**

(ii) links the two hemispheres;

______________________________ **1**

(iii) is rich in the receptor NMDA.

______________________________ **1**

DO NOT WRITE IN THIS MARGIN

Marks

10. The diagram represents a motor neurone.

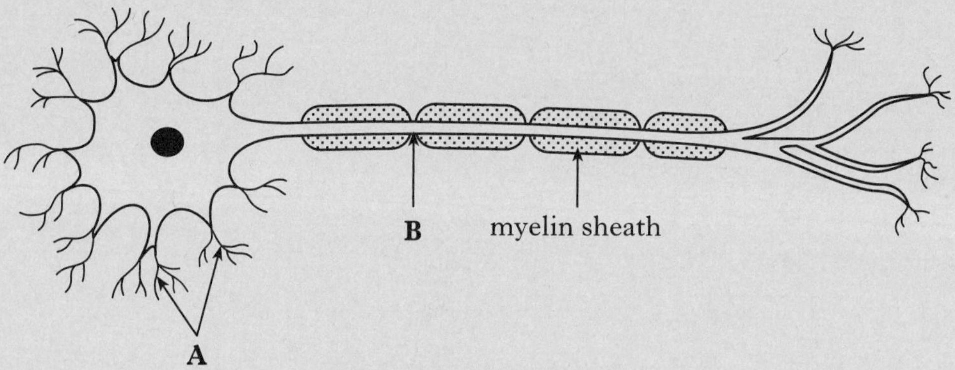

(*a*) Name the nerve fibres **A** and **B**.

A ______________________ **B** ______________________ **1**

(*b*) The table below describes features of somatic and autonomic motor neurone function.

Complete the table.

Feature	*Somatic*	*Autonomic*
type of control (conscious/unconscious)		
example of target muscle		uterine muscle
example of neurotransmitter		noradrenaline

2

(*c*) State the effect of sympathetic stimulation on the:

1 heart rate ______________________

2 digestive system ______________________

3 skin arterioles ______________________ **3**

(*d*) The sympathetic and parasympathetic nervous systems often influence organs in opposite ways. What term describes this opposing effect?

______________________ **1**

DO NOT WRITE IN THIS MARGIN

Marks

11. The diagram shows a simplified outline of the nitrogen cycle.

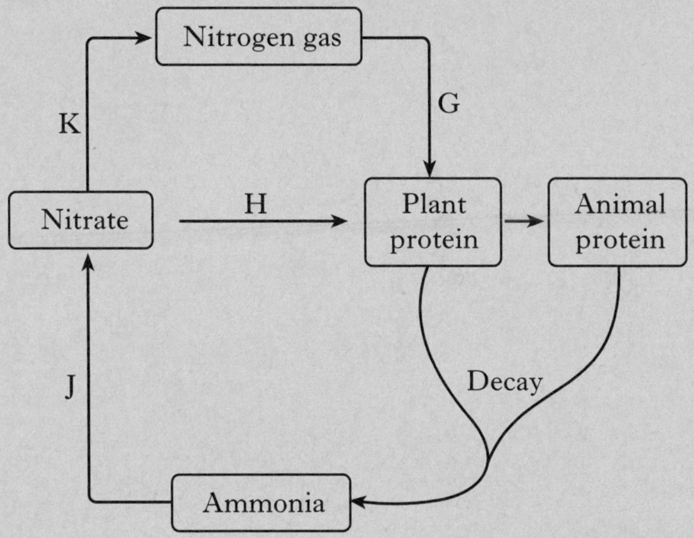

(*a*) The table below shows some of the processes involved in the nitrogen cycle.

Complete the table using information from the diagram.

Label	*Type of bacteria*	*Process in nitrogen cycle*
G		trap atmospheric nitrogen
	nitrifying	
K		convert nitrate to nitrogen gas

2

(*b*) Nitrate is lost from the soil by leaching.

Describe the effect of nitrate pollution on fresh water environments.

____

____

____

____

____

____

____ 3

[Turn over

DO NOT WRITE IN THIS MARGIN

Marks

12. **Graphs 1** and **2** below contain information about measles in Scotland from 1974 to 1999.

Graph 1 – Cases of measles

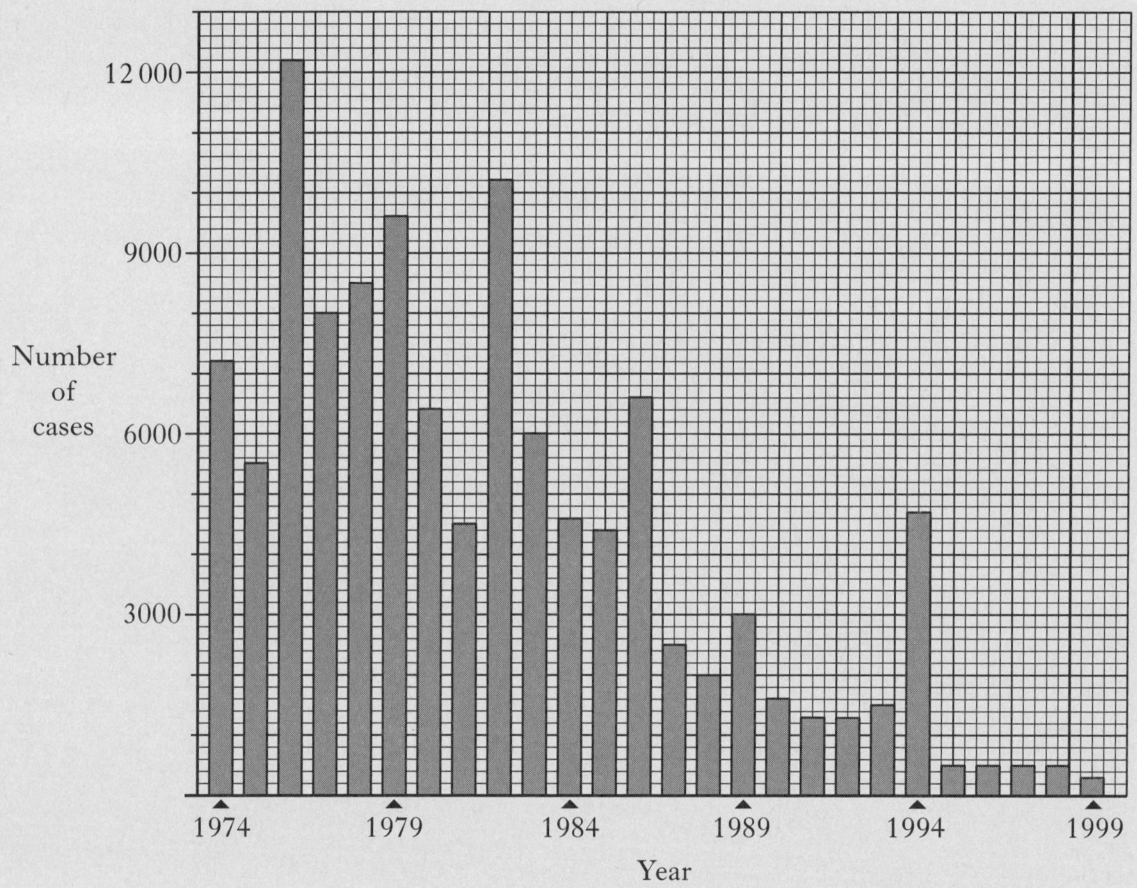

Graph 2 – Deaths from measles

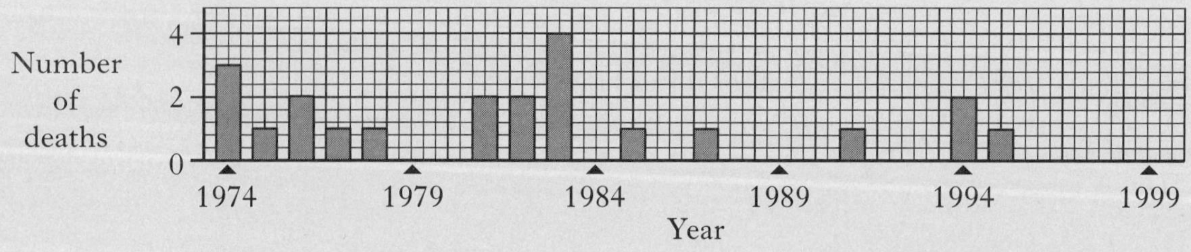

(*a*) From **Graph 1** construct a table to show the years in which measles cases were at their lowest and highest levels. The table should include the number of cases in each year. **2**

DO NOT WRITE IN THIS MARGIN

Marks

12. (continued)

(*b*) Suggest a reason for the trend in measles cases shown in **Graph 1**.

____

____ **1**

(*c*) From **Graphs 1** and **2**, what percentage of individuals who contracted measles in 1995 died from the disease?

Space for calculation

____________% **1**

[Turn over

13. The following information relates to the impact of human activities on the carbon cycle in the year 2000.

Table 1 World carbon reserves

Reservoir	*Mass of carbon stored* (billions of tons of carbon per year)
Oceans	35 000
Fossil fuels	10 000
Soil	1500
Atmosphere	500
Plants	500

Table 2 Mass of carbon released by human activity

Activity	*Mass of carbon released* (billions of tons of carbon per year)
Burning of fossil fuels	5·5
Deforestation	1·5

Table 3 Annual carbon gain by the atmosphere and the oceans

Reservoir	*Mass of carbon gained* (billions of tons of carbon per year)
Atmospheric carbon	3·3
Oceanic carbon	2·0

DO NOT WRITE IN THIS MARGIN

13. (continued) *Marks*

(*a*) Construct a bar chart to illustrate the data in **Table 1**.

(Additional graph paper, if required, can be found on page 28.)

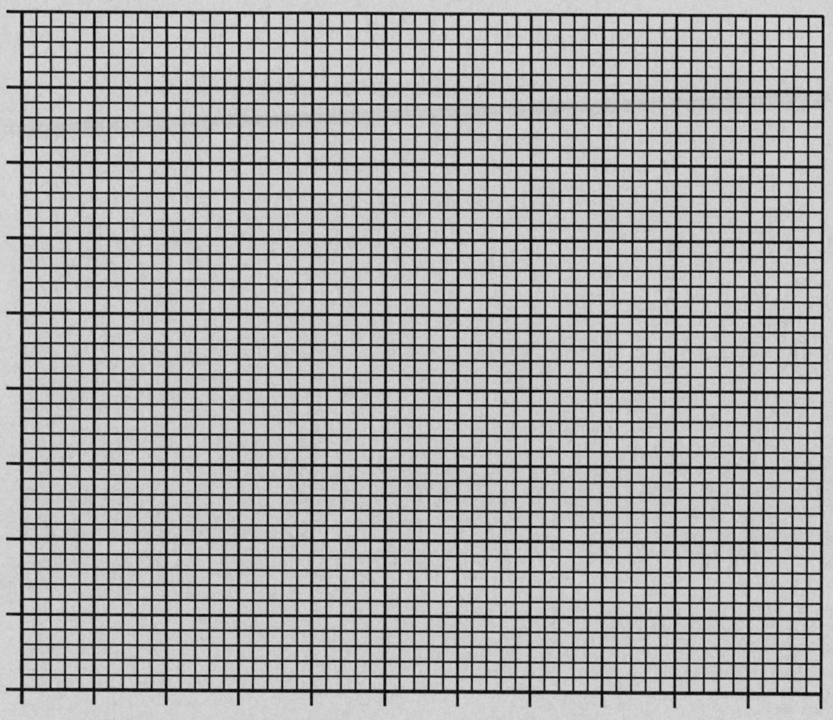

2

(*b*) It is estimated that only 10% of fossil fuel reserves are available for human use. Using information from **Tables 1** and **2** calculate how long these reserves will last.

Space for calculation

__________ 1

(*c*) Use the information from **Tables 1** and **3** to estimate the number of years it will take for atmospheric carbon to exceed 550 billion tons.

Space for calculation

__________ 1

(*d*) State **two** likely consequences of increased atmospheric carbon levels on global climate patterns.

1 ____

2 ____ 2

DO NOT WRITE IN THIS MARGIN

Marks

SECTION C

Both questions in this section should be attempted.

Note that each question contains a choice.

Questions 1 and 2 should be attempted on the blank pages which follow.

Supplementary sheets, if required, may be obtained from the invigilator.

Labelled diagrams may be used where appropriate.

1. Answer **either** A **or** B.

A. Discuss the influence of others on an individual's behaviour under the following headings:

(i) Social facilitation; **3**

(ii) Deindividuation; **3**

(iii) Influences that change beliefs. **4**

(10)

OR

B. Discuss the exponential growth of the human population under the following headings:

(i) Demographic trends; **2**

(ii) Agriculture; **4**

(iii) Disease. **4**

(10)

In question 2 ONE mark is available for coherence and ONE mark is available for relevance.

2. Answer **either** A **or** B.

A. Describe the influence of hormones on the testes. **(10)**

OR

B. Describe the events which take place in the first half of the menstrual cycle. **(10)**

[*END OF QUESTION PAPER*]

DO NOT WRITE IN THIS MARGIN

SPACE FOR ANSWERS

DO NOT WRITE IN THIS MARGIN

SPACE FOR ANSWERS

ADDITIONAL GRAPH PAPER FOR QUESTION 13(*a*)

2004 HIGHER

FOR OFFICIAL USE

Total for Sections B & C

X009/301

NATIONAL QUALIFICATIONS 2004

WEDNESDAY, 19 MAY
1.00 PM – 3.30 PM

HUMAN BIOLOGY HIGHER

Fill in these boxes and read what is printed below.

Full name of centre

Town

Forename(s)

Surname

Date of birth
Day Month Year

Scottish candidate number

Number of seat

SECTION A—Questions 1–30

Instructions for completion of Section A are given on page two.

SECTIONS B AND C

1 (a) All questions should be attempted.

(b) It should be noted that in **Section C** questions 1 and 2 each contain a choice.

2 The questions may be answered in any order but all answers are to be written in the spaces provided in this answer book, and must be written clearly and legibly in ink.

3 Additional space for answers and rough work will be found at the end of the book. If further space is required, supplementary sheets may be obtained from the invigilator and should be inserted inside the **front** cover of this book.

4 The numbers of questions must be clearly inserted with any answers written in the additional space.

5 Rough work, if any should be necessary, should be written in this book and then scored through when the fair copy has been written.

6 Before leaving the examination room you must give this book to the invigilator. If you do not, you may lose all the marks for this paper.

SCOTTISH QUALIFICATIONS AUTHORITY

SECTION A

Read carefully

1 Check that the answer sheet provided is for Human Biology Higher (Section A).

2 Fill in the details required on the answer sheet.

3 In this section a question is answered by indicating the choice A, B, C or D by a stroke made in **ink** in the appropriate place in the answer sheet—see the sample question below.

4 For each question there is only **one** correct answer.

5 Rough working, if required, should be done only on this question paper—or on the rough working sheet provided—**not** on the answer sheet.

6 At the end of the examination the answer sheet for Section A **must** be placed **inside** this answer book.

Sample Question

The digestive enzyme pepsin is most active in the

A mouth

B stomach

C duodenum

D pancreas.

The correct answer is **B**—stomach. A **heavy** vertical line should be drawn joining the two dots in the appropriate box in the column headed **B** as shown in the example on the answer sheet.

If, after you have recorded your answer, you decide that you have made an error and wish to make a change, you should cancel the original answer and put a vertical stroke in the box you now consider to be correct. Thus, if you want to change an answer D to an answer B, your answer sheet would look like this:

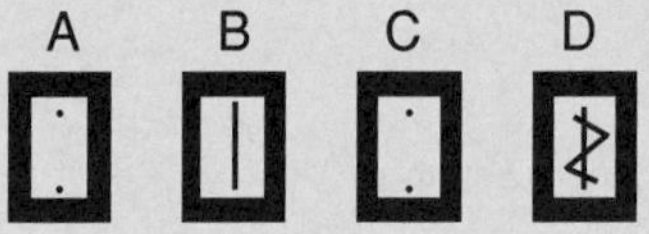

If you want to change back to an answer which has already been scored out, you should enter a tick (✓) to the **right** of the box of your choice, thus:

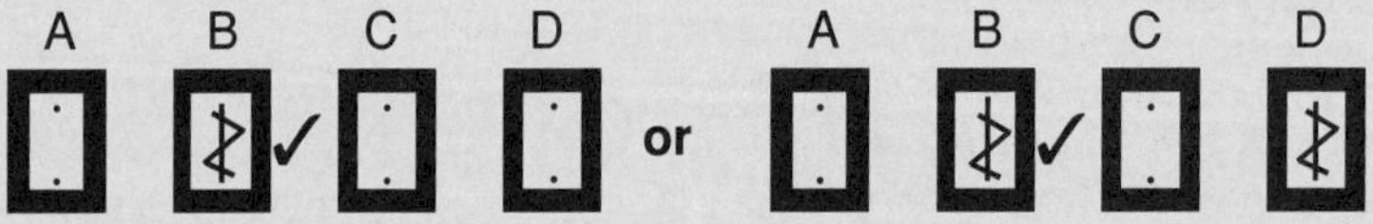

SECTION A

All questions in this section should be attempted.

Answers should be given on the separate answer sheet provided.

1. In respiration, the sequence of reactions resulting in the conversion of glucose to pyruvic acid is called

A the Krebs cycle

B the citric acid cycle

C glycolysis

D the cytochrome chain.

2. Which of the following is an insoluble polysaccharide?

A Glycogen

B Glucose

C Sucrose

D Maltose

3. Which of the following is **not** a function of lipids?

A Nerve insulation

B Vitamin transport

C Energy storage

D Oxygen transport

4. Which of the following processes requires infolding of the cell membrane?

A Diffusion

B Phagocytosis

C Active transport

D Osmosis

5. The formation of new viruses involves the following stages:

X viral protein coats are synthesised

Y host cell metabolism is taken over by virus

Z viral nucleic acid is replicated.

The correct order in which these stages occur is

A X → Z → Y

B Y → X → Z

C Z → X → Y

D Y → Z → X.

6. A sex-linked gene carried on the X-chromosome of a man will be transmitted to

A 50% of his male children

B 50% of his female children

C 100% of his male children

D 100% of his female children.

7. The family tree shows the inheritance of red-green colour blindness in humans. Red-green colour blindness is a recessive, sex-linked condition.

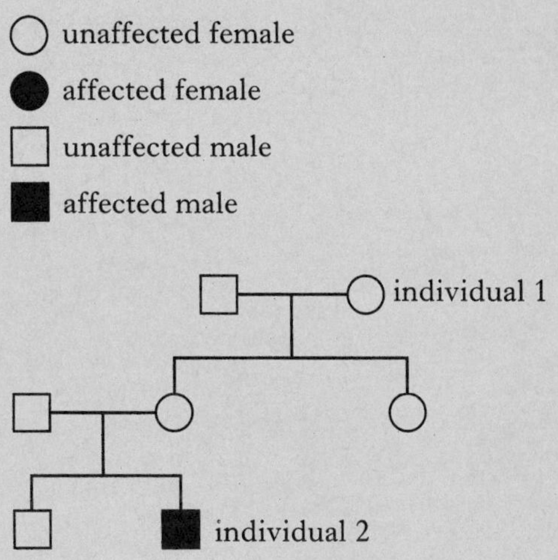

Which line in the table describes correctly the genotypes of individual 1 and individual 2?

	Individual 1	*Individual 2*
A	X^RX^R	X^RY
B	X^RX^r	X^RY
C	X^rX^r	X^rY
D	X^RX^r	X^rY

[Turn over

8. Which of the following describes the term non-disjunction?

A The failure of chromosomes to separate at meiosis

B The independent assortment of chromosomes at meiosis

C The exchange of genetic information at chiasmata

D An error in the replication of DNA before cell division

9. The diagram below shows a cross-section of a testis.

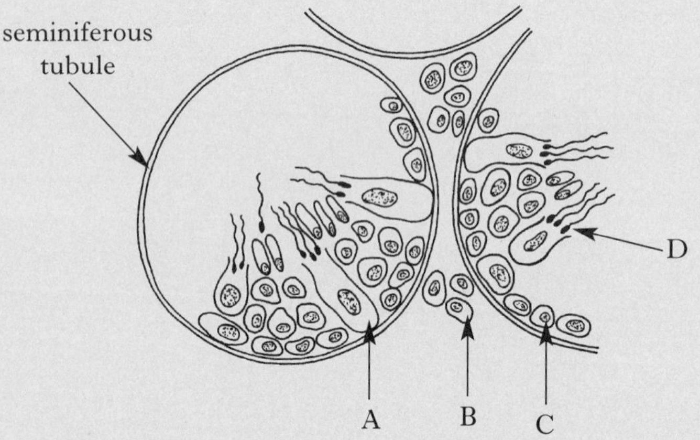

Which cell can produce testosterone?

10. The graph below shows the growth, in length, of a human fetus before birth.

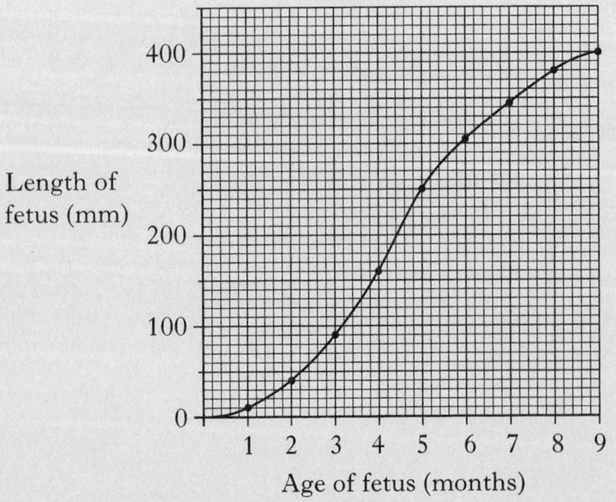

What is the percentage increase in length of the fetus during the final 4 months of pregnancy?

A 33·3

B 60·0

C 62·5

D 150·0

11. The graph below shows the dissociation curves for fetal and maternal haemoglobin.

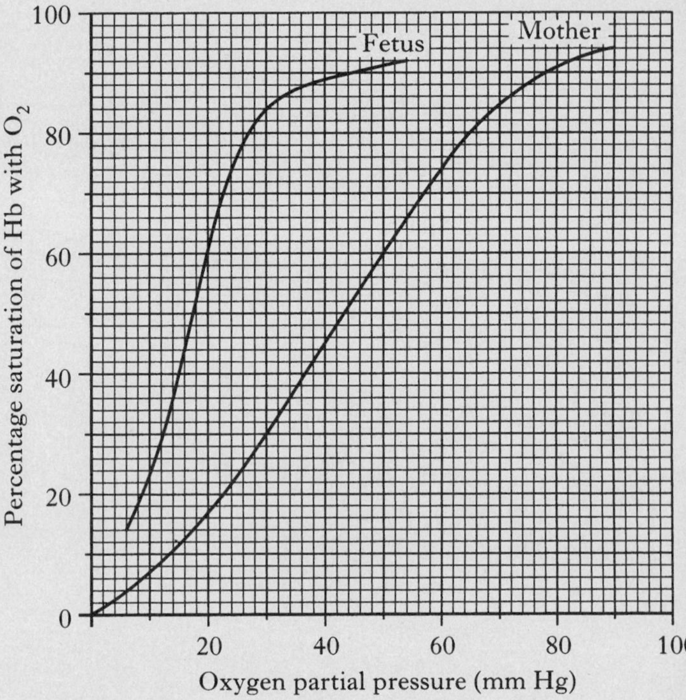

What is the difference in percentage saturation of haemoglobin between the mother and the fetus at a partial pressure of 30 mm Hg?

A 18

B 19

C 52

D 54

12. Which of the following are required for red blood cell production?

A Iron and vitamin D

B Calcium and vitamin B_{12}

C Iron and vitamin B_{12}

D Calcium and vitamin D

13. Colostrum provides a baby with

A antibodies

B antigens

C phagocytes

D lymphocytes.

14. The graph shows changes in lung volume during a breathing exercise.

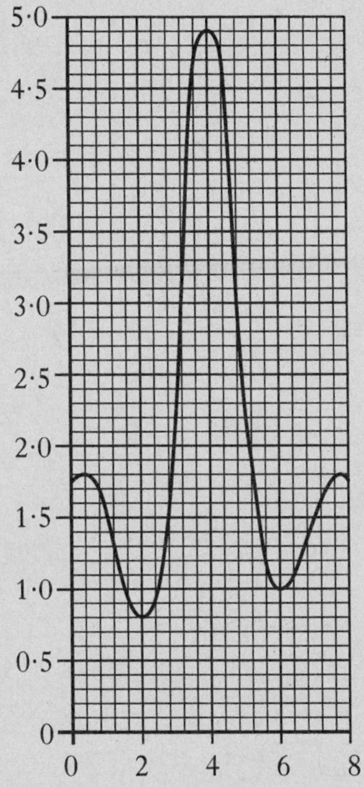

What is the volume of air inhaled between 2 and 4 seconds?

A 0·8 litres

B 3·9 litres

C 4·1 litres

D 4·9 litres

15. Which **two** blood vessels are involved in the transport of blood to and from the head?

A Carotid artery and jugular vein

B Renal artery and pulmonary vein

C Aorta and renal vein

D Hepatic artery and jugular vein

16. The table below shows the relative concentrations of certain substances in blood vessels leading to and from the liver.

(+++ = high, ++ = moderate, + = low)

Blood vessel	*Oxygen*	*Carbon dioxide*	*Urea*	*Amino acids*
1	+++	+	+	+
2	+	+++	+	+++
3	+	+++	+++	+

Which line of the table below identifies correctly the blood vessels?

	Hepatic vein	*Hepatic portal vein*	*Hepatic artery*
A	1	2	3
B	2	3	1
C	3	2	1
D	3	1	2

17. Which line of the table identifies correctly the hormones which stimulate the inter-conversion of glucose and glycogen?

	glucose → glycogen	*glycogen → glucose*
A	insulin	glucagon and adrenaline
B	glucagon and insulin	adrenaline
C	adrenaline and glucagon	insulin
D	adrenaline	glucagon and insulin

[Turn over

18. Which of the following shows the substance from which urea is produced and the site of urea production?

	Substance	*Site of production*
A	amino acid	liver
B	amino acid	kidney
C	glycogen	liver
D	glycogen	kidney

19. What is the function of the glomerulus in the production of urine?

A Collection of filtrate

B Filtration of blood

C Reabsorption of glucose

D Osmoregulation

20. The concentration of urea in plasma and urine is given in the table below.

	Plasma	*Urine*
Urea (g/100 cm^3)	0·3	1·29

By how many times has the urea been concentrated by the activity of the kidney?

A 0·23 times

B 0·39 times

C 4·3 times

D 43 times

21. The diagram below shows a body shape made up of cubes.

The surface area to volume ratio of this body is

A 4 : 1

B 6 : 1

C 15 : 4

D 29 : 8

22. The temperature monitoring centre of the brain is in the

A medulla oblongata

B cerebellum

C pituitary gland

D hypothalamus.

23. The diagram below shows the body's response to temperature change.

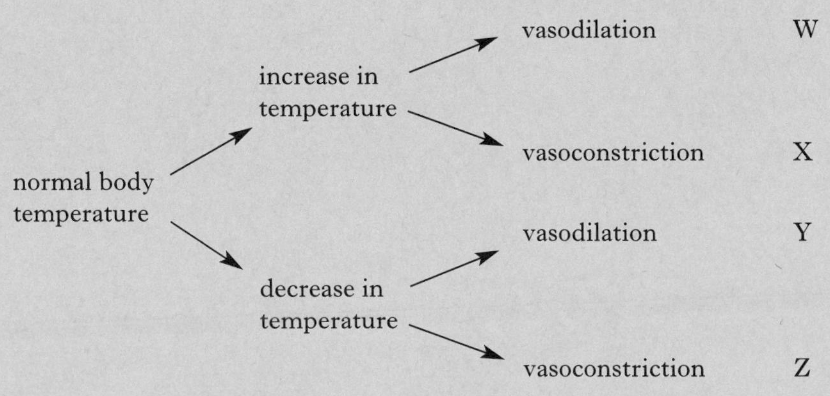

Which letters indicate negative feedback control of body temperature?

A W and Y

B W and Z

C X and Y

D X and Z

---

24. The peripheral nervous system contains the

A brain and spinal cord

B brain and somatic system

C spinal cord and autonomic system

D somatic system and autonomic system.

25. An investigation was carried out to determine how long it takes a student to learn the pathway through a finger maze. The student was allowed to complete the maze ten times. Which of the following pairs of factors would have to be kept the same each time?

A The time taken to complete the maze and the shape of the maze

B The number of errors made and the finger used

C The finger used and the shape of the maze

D The time taken to complete the maze and the finger used

26. Which of the following best describes social facilitation?

A Improved performance in the presence of others

B Deindividuation in competitive situations

C Discrimination behaviour shown by groups of individuals

D Shaping behaviour as seen in trial and error learning

27. Why do humans have a long period of dependency?

A To allow for learning and the development of language

B To allow bonding to take place between mother and child

C To allow for the learning of motor and sensory skills

D To allow for the growth of the brain and other major body organs

28. The diagram below shows the carbon cycle.

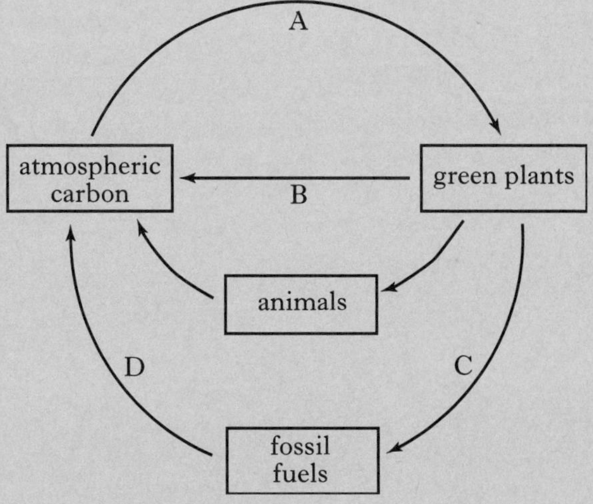

Which letter represents respiration?

[Turn over

29. In the nitrogen cycle, which of the following processes is carried out by nitrifying bacteria?

The conversion of

A nitrate to ammonia

B ammonia to nitrate

C nitrogen gas to ammonia

D nitrogen gas to nitrate.

30. An algal bloom in a loch may result from

A lack of oxygen

B lack of sunlight

C excess phosphates

D excess herbicide.

Candidates are reminded that the answer sheet MUST be returned INSIDE the front cover of this answer booklet.

DO NOT WRITE IN THIS MARGIN

Marks

SECTION B

All questions in this section should be attempted.

1. The diagram shows the role of ATP in cell metabolism.

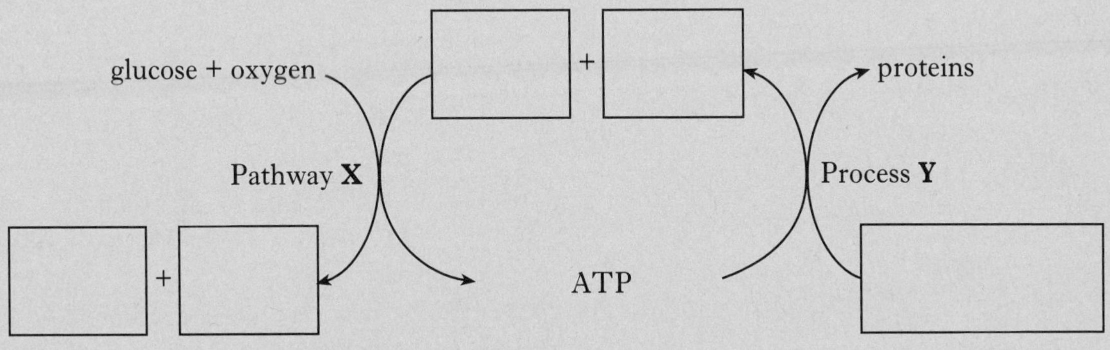

(*a*) Complete the diagram by entering the names of the appropriate substances. **3**

(*b*) (i) Name **one** stage of pathway **X** and state where it occurs in the cell.

Stage ____________________ Location ____________________ **1**

(ii) Name the organelle where process **Y** occurs.

______________________________ **1**

(*c*) Describe **two** ways in which the diagram would be different under anaerobic conditions.

1 ____

____

2 ____

____ **2**

(*d*) Name a respiratory substrate other than glucose.

______________________________ **1**

[Turn over

DO NOT WRITE IN THIS MARGIN

Marks

2. Sickle-cell anaemia is a blood disorder in which haemoglobin is malformed.

The diagram below shows the effect of this disorder on a red blood cell.

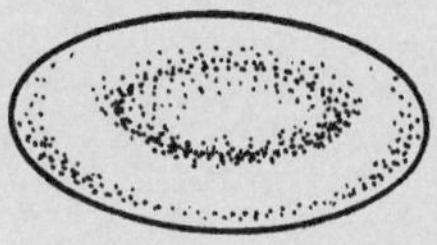

Normal red blood cell

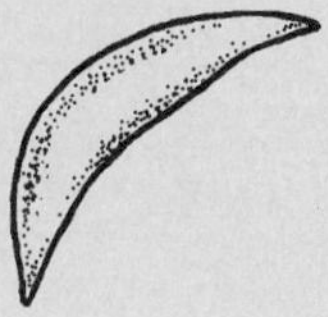

Sickled red blood cell

The condition is not sex-linked. The allele for normal haemoglobin (**H**) is incompletely dominant to the sickle-cell allele (**h**).

Heterozygous individuals are mildly affected, whereas those with genotype **hh** are severely affected.

Two mildly affected parents have two children who are mildly affected like their parents. The parents are expecting a third child.

(*a*) Complete the Punnett square to show the possible genotypes of this child.

2

(*b*) From the Punnett square calculate the percentage chance of the child being

1 unaffected __________

2 mildly affected __________

3 severely affected. __________ 1

DO NOT WRITE IN THIS MARGIN

Marks

3. The diagram below shows a section through a lymph node.

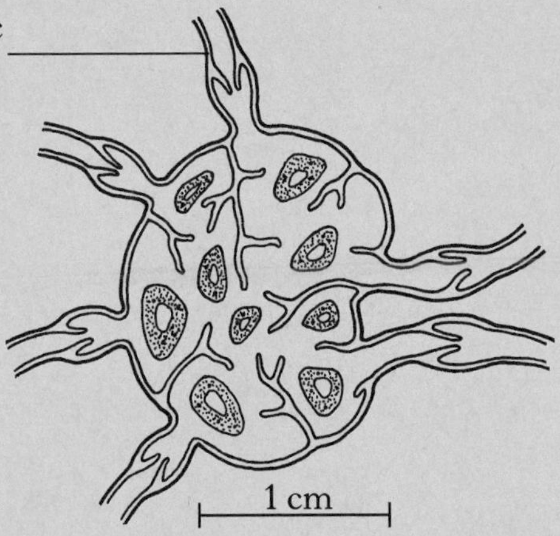

(*a*) Complete the table to name the cells found in the node, and to describe their functions.

Type of cell	*Secretion of antibodies (yes/no)*	*Type of response*
B-lymphocyte		
	no	cell-mediated response
		non-specific response

3

(*b*) Add an arrow to the diagram to indicate the direction of flow of lymph in vessel X. Give a reason for your choice.

Reason ______________________________

______________________________ 1

(*c*) Describe **one** way in which the composition of lymph differs from plasma.

____

______________________________ 1

(*d*) What eventually happens to the lymph after it leaves the gland?

______________________________ 1

(*e*) Describe **one** function of the lymphatic system, apart from protecting the body from infection.

____

______________________________ 1

DO NOT WRITE IN THIS MARGIN

Marks

4. An investigation was carried out into the effect of caffeine on blood pressure, using coffee as the source of caffeine.

The systolic and diastolic blood pressures of six students were measured using a digital sphygmomanometer. Each student was then given a cup of coffee to drink. After one hour their blood pressure was measured again.

The results are shown in the table below.

Student	*Initial blood pressure* (mmHg)	*Final blood pressure* (mmHg)
1	120/75	134/82
2	127/79	145/88
3	118/70	124/72
4	134/81	143/83
5	122/73	133/77
6	129/84	137/90
Average	**125/77**	

(*a*) Calculate the average final blood pressure and write your answer in the table above.

Space for calculation

1

(*b*) What conclusion can be drawn from these results?

____

____ **1**

(*c*) Describe an appropriate control for this investigation.

____

____ **1**

(*d*) Apart from leaving one hour between readings, list **two** other variables which would need to be kept constant during this investigation.

1 ____

2 ____ **1**

DO NOT WRITE IN THIS MARGIN

4. (continued) *Marks*

(*e*) What is meant by systolic and diastolic blood pressure?

Systolic ____

____

Diastolic ____

____ **2**

(*f*) The graphs below show initial and final blood pressures of one of the students.

Graph 1 Initial Blood Pressure

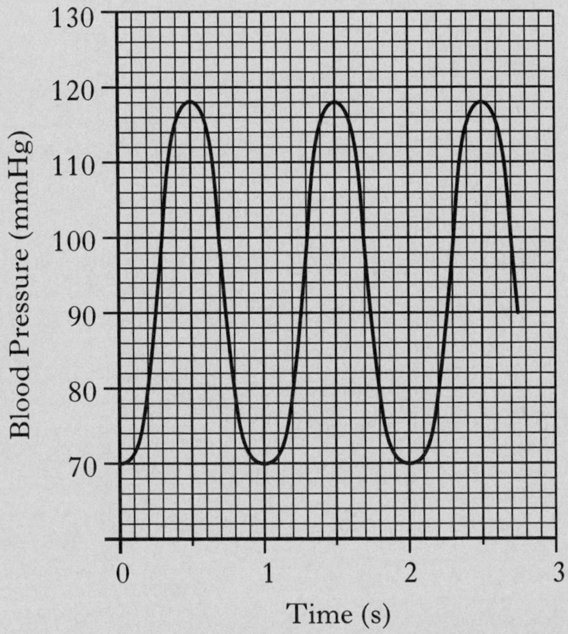

Graph 2 Final Blood Pressure

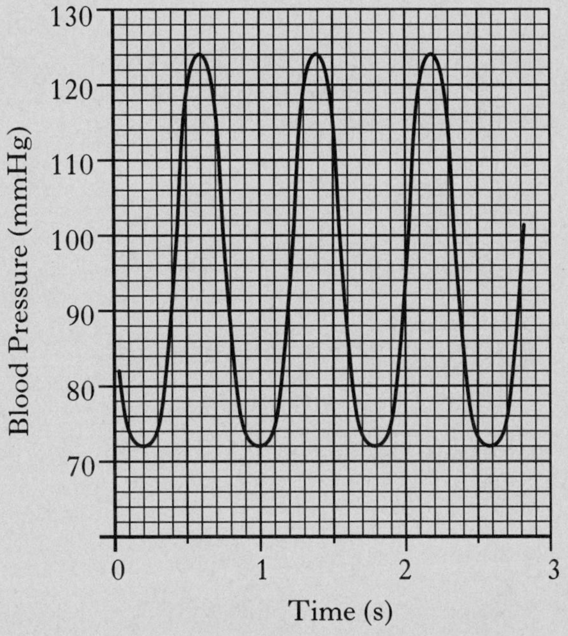

(i) Use the information in the table and the graphs to identify the student.

Student number ____________ **1**

(ii) Calculate the increase in the pulse rate of this student over the period of the investigation.

Space for calculation

__________ bpm **1**

[Turn over

DO NOT WRITE IN THIS MARGIN

Marks

5. The diagram shows stages in the development of a human embryo from fertilisation to implantation.

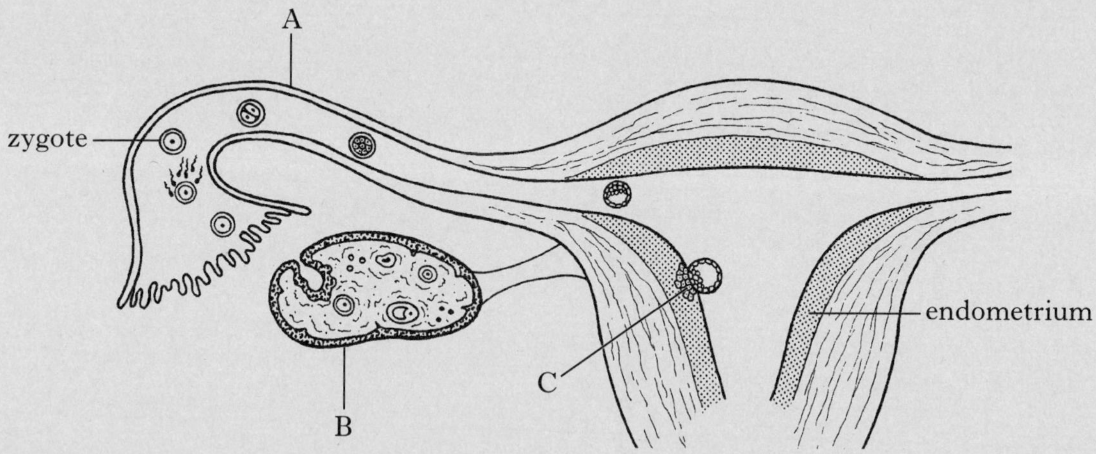

(*a*) Name the parts labelled **A** and **B**.

A ______________________ B ______________________ 1

(*b*) What term is used to describe the first few divisions of the zygote?

______________________ 1

(*c*) Name a hormone which is involved in preparing the endometrium for implantation and state where it is produced.

Hormone ______________________ Produced by ______________________ 1

(*d*) What organ will develop from the tissue labelled **C**?

______________________ 1

(*e*) Sometimes twins develop in the uterus. Distinguish between the formation of monozygotic and dizygotic twins.

____

____

____

____

____

____

______________________ 3

DO NOT WRITE IN THIS MARGIN

Marks

6. The sperm counts of 30 men taken between 1940 and 2000 are shown in the graph below. A line of best-fit has been drawn, to indicate the trend over the 60-year period.

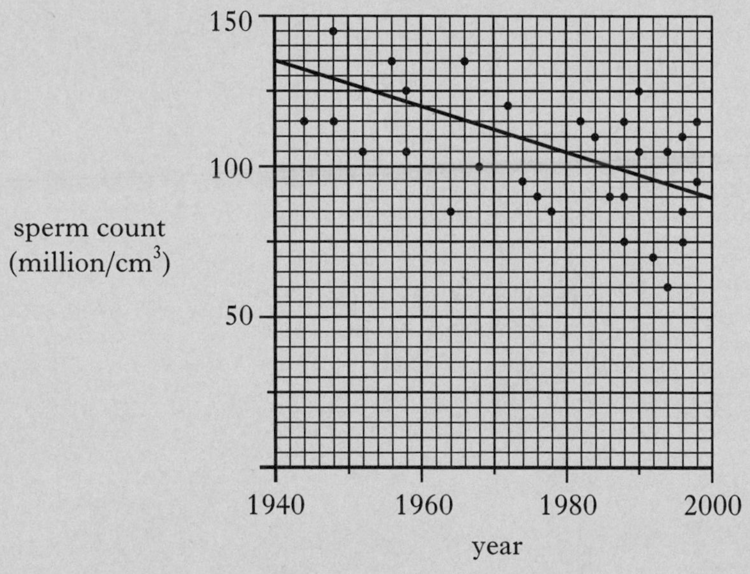

(*a*) Using the line of best-fit, calculate the percentage decline in sperm count over the 60-year period.

Space for calculation

_____________ % **1**

(*b*) From the graph, what is the maximum sperm count for any one individual recorded during this period?

_____________ million/cm^3 **1**

(*c*) Some insecticides are thought to influence sperm production. Explain why animals at the end of food chains are more likely to be affected by insecticides.

____

____

____ **1**

(*d*) Name the pituitary hormone which stimulates the production of sperm.

________________________ **1**

[Turn over

DO NOT WRITE IN THIS MARGIN

Marks

6. (continued)

(*e*) Name a gland which adds fluid to sperm during ejaculation and describe **one** function of this fluid.

Gland ______________________ 1

Function of fluid ______________________

______________________ 1

(*f*) Two treatments sometimes used for infertility are artificial insemination and *in vitro* fertilisation. Describe briefly what is meant by these terms.

artificial insemination ______________________

____

in vitro fertilisation ______________________

______________________ 2

DO NOT WRITE IN THIS MARGIN

Marks

7. The Rhesus blood group system is determined by three pairs of alleles: **Cc**, **Dd** and **Ee**. However, only the **D** allele is important in blood transfusion and pregnancy. People with the dominant allele **D** are Rhesus positive and those with genotype **dd** are Rhesus negative.

(*a*) What term is used to describe characteristics controlled by many pairs of alleles?

______________________________ **1**

(*b*) Name another blood group system which has to be matched for blood transfusion to be successful.

______________________________ **1**

(*c*) What part of a cell carries the Rhesus antigen?

______________________________ **1**

(*d*) A Rhesus negative woman and a Rhesus positive man are planning to have a child. They consult a genetics counsellor to find out whether their child is likely to be Rhesus positive or Rhesus negative.

What genetic information could they be given?

____

____

____

____

____

____ **2**

(*e*) Describe a treatment which can be used to protect a child at risk from the Rhesus reaction.

____

____

____ **1**

[Turn over

DO NOT WRITE IN THIS MARGIN

Marks

8. The table below gives data on kidney transplants for the UK in the year 2001.

Category of patient	*Number*
On waiting list at beginning of 2001	**6052**
Received transplants during the year	
Removed from the list during the year	293
Died during the year	203
Added to the list during the year	1921
On waiting list at end of 2001	**6241**

(*a*) Complete the table by calculating the number of patients who received a transplant during the year 2001.

Space for calculation

1

(*b*) Assuming the same trend for the year 2002, predict the number of patients on the waiting list at the end of that year.

Space for calculation

__________ 1

(*c*) Drugs which suppress the immune system are given to transplant patients. Explain how this treatment reduces the chance of rejection of the transplanted kidney.

____

____

____

______________________________ 2

(*d*) Name the **two** blood vessels which would have to be cut and reconnected during a kidney operation.

_______________ and _______________ 1

DO NOT WRITE IN THIS MARGIN

Marks

9. The diagram below is of a motor homunculus which represents the relative sizes of parts of the brain associated with motor control.

(*a*) In which part of the brain is the motor area located?

______________________________ **1**

(*b*) What is the function of the motor area of the brain?

______________________________ **1**

(*c*) Explain why the hands have such a large area of the brain devoted to their control in comparison to the feet.

____

______________________________ **1**

(*d*) What type of neural pathway is used to co-ordinate movements of the fingers?

______________________________ **1**

(*e*) Three facial expressions are shown below.

What term describes this type of communication? **1**

____

DO NOT WRITE IN THIS MARGIN

Marks

10. The diagram shows a neuromuscular junction.

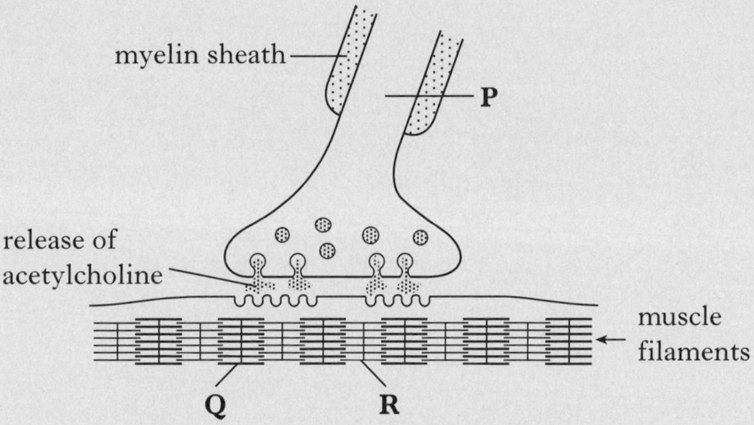

(*a*) Name the part of the nerve cell labelled **P**.

______________________________ 1

(*b*) (i) What kind of substance is acetylcholine?

______________________________ 1

(ii) What triggers the release of acetylcholine?

____

______________________________ 1

(iii) State what happens to acetylcholine after it has acted on the muscle.

____

______________________________ 1

(*c*) Name the **two** muscle proteins labelled **Q** and **R**.

Q ______________________ **R** ______________________ 1

(*d*) Describe what happens to these protein filaments when a muscle contracts.

____

______________________________ 1

DO NOT WRITE IN THIS MARGIN

Marks

11. Cod have been caught off the coast of Scotland for many years. The graph below shows the estimated population of cod in an area of the North Sea over the last hundred years.

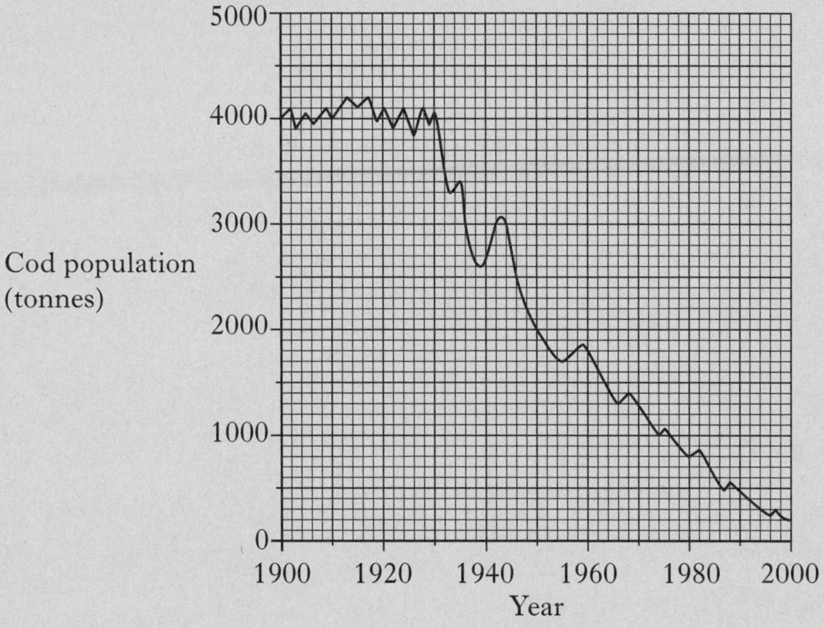

(*a*) Between the years 1900 and 1930 this area of the North Sea was at its carrying capacity for the cod population.

Explain what is meant by the term "carrying capacity".

____

____ **1**

(*b*) (i) Express, as a simple whole number ratio, the size of the cod population in 1950 to its size in 2000.

Space for calculation

______ : ______ **1**

(ii) Suggest **two** reasons for the decline of the cod population over the last 50 years.

1 ____

____

2 ____

____ **2**

[Turn over

DO NOT WRITE IN THIS MARGIN

Marks

12. The graphs show the changes in birth and death rates of two countries **A** and **B**.

Country A

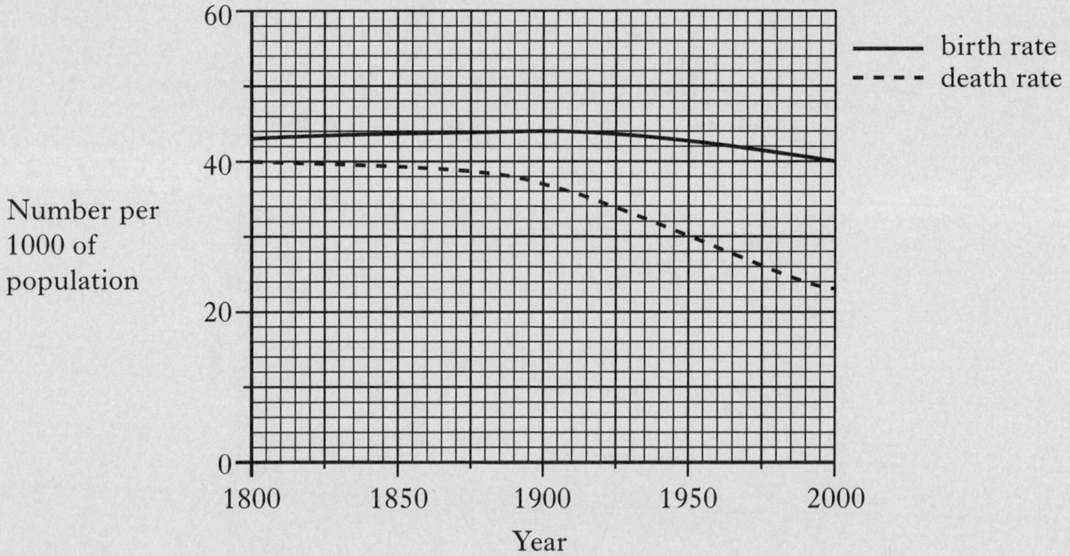

Country B

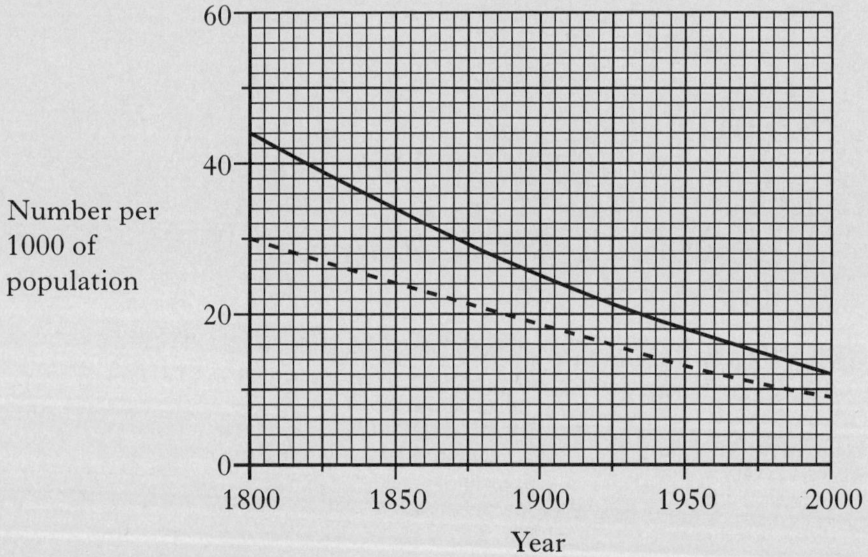

(*a*) (i) State the birth rate and death rate in country **A** in the year 1900.

Birth rate ______________ Death rate ______________ number per 1000 **1**

(ii) Suggest **one** reason for the decline in birth rate of country **A** over the last fifty years.

____

______________________________ **1**

(iii) Suggest **one** reason for the decline in death rate of country **A** over the last fifty years.

____

______________________________ **1**

DO NOT WRITE IN THIS MARGIN

Marks

12. (continued)

(*b*) Is the population of country **B** increasing or decreasing over the period of time shown?

Give a reason for your answer.

Change ______________________

Reason ______________________

____

______________________ 1

(*c*) Which of the **two** countries is likely to be a developing country?

Give a reason for your answer.

Country ______________________

Reason ______________________

____

______________________ 1

[Turn over

DO NOT WRITE IN THIS MARGIN

Marks

13. An investigation was carried out into the effect of lack of nitrates on the growth of cress seedlings. Two equal batches of seeds were grown in agar gel containing:

A all necessary mineral salts

B all mineral salts except nitrate.

Each seed was placed on the surface of the agar in a glass tube as shown in the diagram below.

The heights of the seedlings were measured every day for eight days and the results are shown in the table.

Day	*Average height of seedlings in gel A* (mm)	*Average height of seedlings in gel B* (mm)
0	0	0
1	1	1
2	3	3
3	6	4
4	9	5
5	13	6
6	18	7
7	23	8
8	29	8

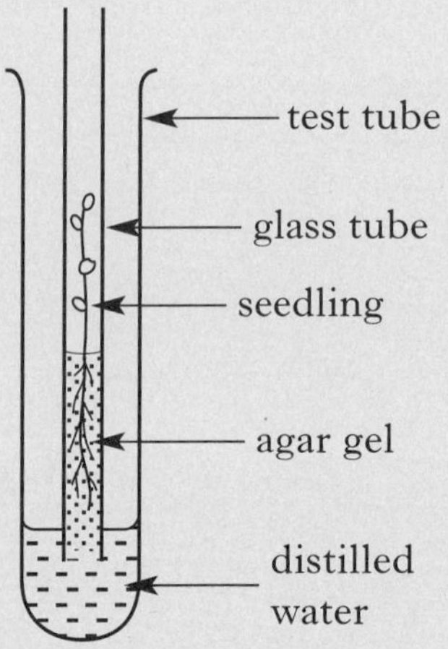

(*a*) Construct a line graph to illustrate the data in the table.

(Additional graph paper, if required, can be found on page 32.)

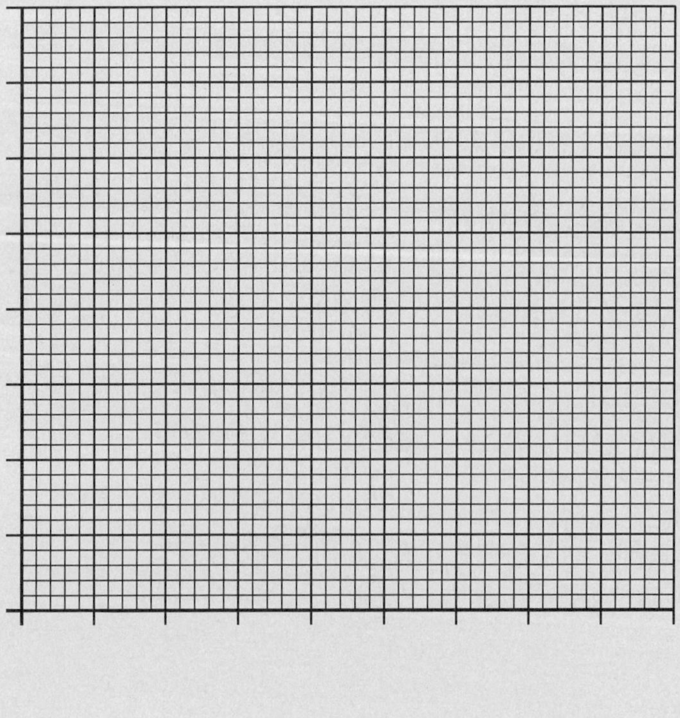

3

(*b*) State **one** conclusion which can be drawn from these results.

1

DO NOT WRITE IN THIS MARGIN

13. (continued) *Marks*

(*c*) Identify the control and explain your choice.

Control ______________________

Explanation ____

____ 1

(*d*) What feature of this investigation makes the results reliable?

____ 1

(*e*) Suggest why distilled water is used in the test-tube rather than tap water.

____

____ 1

(*f*) Phosphates are also necessary for good plant growth. Name **one** compound, other than ADP and ATP, which contains phosphate.

______________________________ 1

[Turn over

DO NOT WRITE IN THIS MARGIN

Marks

SECTION C

Both questions in this section should be attempted.

Note that each question contains a choice.

Questions 1 and 2 should be attempted on the blank pages which follow.

Supplementary sheets, if required, may be obtained from the invigilator.

Labelled diagrams may be used where appropriate.

1. Answer **either** A **or** B.

A. Give an account of memory under the following headings:

(i) short term memory; **4**

(ii) methods of transfer to long term memory. **6**

(10)

OR

B. Describe ways in which food production has been increased in the last fifty years under the following headings:

(i) land use; **4**

(ii) the use of chemicals. **6**

(10)

In question 2 ONE mark is available for coherence and ONE mark is available for relevance.

2. Answer **either** A **or** B.

A. Describe the events in meiosis which give rise to variation in gametes. **(10)**

OR

B. Describe how proteins are assembled from the code on a mRNA strand. **(10)**

[*END OF QUESTION PAPER*]

DO NOT WRITE IN THIS MARGIN

SPACE FOR ANSWERS

DO NOT WRITE IN THIS MARGIN

SPACE FOR ANSWERS

DO NOT WRITE IN THIS MARGIN

SPACE FOR ANSWERS

DO NOT WRITE IN THIS MARGIN

SPACE FOR ANSWERS

DO NOT WRITE IN THIS MARGIN

SPACE FOR ANSWERS

DO NOT WRITE IN THIS MARGIN

SPACE FOR ANSWERS

ADDITIONAL GRAPH PAPER FOR QUESTION 13(*a*)

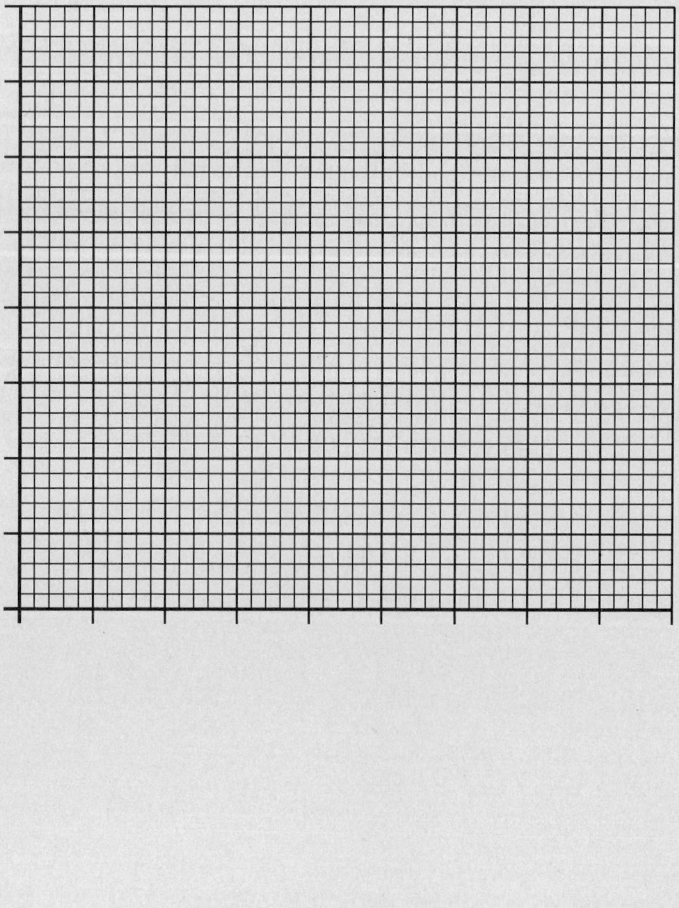

[BLANK PAGE]

[BLANK PAGE]

100 GREAT MOMENTS IN NEW ZEALAND RUGBY

100 GREAT MOMENTS IN NEW ZEALAND RUGBY

TONY WILLIAMS

The front cover shows, clockwise from top left, George Nepia (Peter Bush photo), Sean Fitzpatrick (New Zealand Herald photo), Grant Batty (Peter Bush photo) and the Sopers (family photo).

David Ling Publishing Limited
PO Box 34-601, Birkenhead
Auckland 10, New Zealand

100 Great Moments in New Zealand Rugby
First Edition

ISBN 0-908990-58-8

First Published 1999

Typeset by ExPress Communications Limited
Printed in Australia

CONTENTS

INTRODUCTION

In this book a "moment" is a rugby happening that made time stand still. No modern player captures that better than Jeff Wilson. After he has scored a spectacular (and sometimes unexpected) try his face is momentarily frozen in an expression of both joy and disbelief. For that split second the clock takes a breather, the TV commentator pauses, the players on the field cease their battle and the crowd releases all its tension. Time has come to a shuddering halt in respectful awe of the preceding events.

Often the image of the moment (or a split second of part of it) is frozen by the lens of the cameraman. In this book the moments are frozen in words that recapture and relive one hundred of the greatest moments of New Zealand rugby. Here "great" means "remarkable" and "outstanding". This is not an unending catalogue of victories, but an eclectic and highly personalised choice of those moments that linger in the memory, punctuating the history of New Zealand rugby.

Too often we get lost in statistics and who has scored the most points. Ask a player about his 50th try and he'll give you a blank look, but he'll happily tell you about the one where the pressure was on and he beat three players and dived through the legs of the last to score.

There is a broad mix of incidents listed here ranging from the historical to the bizarre, from crowded international stadiums to country paddocks. What all these stories have in common is that they contain a ball of oval leather, have a New Zealand involvement and they each in their own way communicate the true spirit of rugby in this country.

The stories are told chronologically and that often means the same names keep cropping up during the same era. Though rugby is very much a team sport it is the courage or skill of one individual that is often the most memorable. Giants of the game such as Maurice Brownlie in the 30s, Colin Meads in the 60s and Lomu and Cullen in the 90s have all left their indelible stamp upon the era in which they played.

This book is an ode to New Zealand rugby. It intends to evoke rather than record; to entertain rather than be comprehensive. At times it probably runs the risk of upsetting sober-minded rugby purists. Neither the author nor the publisher take any responsibility for that. Rugby is there to be enjoyed. So enjoy, and wallow in the great moments of our rugby folklore.

ACKNOWLEDGEMENTS

As I have not personally witnessed all the incidents related here, I am grateful to others who have recorded matches of the past. But of course the greatest debt is to those who have played the game and thrilled us all with so many memorable rugby moments.

1

THE FIRST GAME

4 May 1870. The Botanics Reserve, Nelson

The day when football in New Zealand first became rugby

Rugby is supposed to have derived from Rugby School where an errant schoolboy William Webb Ellis was said to have first run with the ball. At the time there were many versions of "football" being played, but it was the handling version played at Rugby School which became universally adopted, mainly it seems, because the school was the first to publish its rules – on 28 August 1845.

Various forms of "rugby" were played long before that. A report of one such match which took place in England in 1763 definitely sounds like rugby. It stated, "...though many were injured, fortunately none were killed." Games similar to rugby were also played in antiquity as artwork that has come down from the Aztecs, Greeks, Romans and Chinese attest.

The first official rugby match in New Zealand is generally considered to have taken place on 14 May 1870. Prior to that "football" (soccer) and Victorian rules (later renamed Australian rules) were generally played. The story goes that Charles Monro, son of Sir David Monro, then Speaker of the New Zealand Parliament was sent to the English public school, Christ College, Finchley to further his education. In doing so he furthered New Zealand's education when he arrived back in New Zealand on the *Airedale* on 27 January 1870 bringing with him details and rules of the game he had so enjoyed playing in England.

By 14 May he had interested sufficient numbers of his contemporaries in the new sport to engage in a contest at the Botanics Reserve in Nelson between Nelson Town and Nelson College. Monro played for the Nelson team and the College team was selected from the 61 students on the roll.

There were 18 players on each side (a number agreed to by the captains before the match). The teams consisted of 10 forwards, three half backs, three three-quarters and two fullbacks. About 200 spectators were in attendance, including "a goodly number of the opposite sex who seemed to enjoy the changeful fortunes of the game", as well as Charles's parents Sir David and Lady Monro.

College won the toss and elected to kick off. A report of the match in the *Nelson Colonist* of 17 May 1870 stated, " ...And now what was comparative silence and inactivity is suddenly converted into a rushing, noisy, shouting crew, and as

"it is all shove, pull, rush and roll about in a confused mass..."

Nelson Colonist
17 May 1870

the ball is kicked off the game commences in earnest. Almost as soon as the ball is amongst the Town players, the College have followed up, when a kick sends it over their heads again, and then it is seen in one place, and then in another, the whole field in pursuit. Now some player runs with it, and a general scrimmage ensues; it is all shove, pull, rush and roll about in a confused mass until 'down' is cried, and away the ball goes again till perchance it gets in touch or caught."

Nelson Town won by 2 goals to nil. The first game of rugby had been played in New Zealand, but it seems no one was aware of it yet as the local newspapers reported it as just another "football" match.

2

THE GOOD OLD DAYS

3 October 1882. The Domain, Auckland

New South Wales were the first overseas rugby team to tour New Zealand

"...there were no fists thumped into palms, no team talks and no desperate exhortations from coaches. Instead the two teams mixed freely and chatted to each other..."

It is hard not to be nostalgic about the early days of rugby before the All Blacks even existed or had acquired the desperate need to win every match they played. In 1881 the game was sufficiently organised in this country to consider an invitation from the Southern Rugby Union (renamed the New South Wales Rugby Union the following year) to tour Australia.

The secretary of the Wellington Rugby Union tried to arouse interest in New Zealand, but when he found little of it, he instead invited the New South Welshmen to come here. The invitation was duly accepted and New South Wales became the first overseas rugby team to tour New Zealand.

The team arrived in Auckland on 6 September with just 16 players. They had an inauspicious start to the tour when their boat the *Rotomahana* ran aground and they had to complete their journey onto New Zealand soil on the Devonport ferry.

New South Wales played seven matches throughout the country. They lost to a grouping called Auckland Provincial Clubs, and to Otago, but were able to beat Wellington (on the way south), West Coast (played at Wellington), Canterbury and Wellington again (on the way north).

The final match of the tour was again against Auckland Provincial Clubs. Many local businesses decided to declare an unofficial holiday, closing down for the day and hurrying to the match which was attended by 3,000 people. It was the local custom to play matches in four "quarters" of half an hour each, but in deference to the visitors the match was played with two 45-minute halves. At the start of the match New South Wales decided to play with 10 of their players as forwards.

Due to the rain that fell throughout most of the match there was much dribbling of the ball. Auckland scored the first try when Whiteside went on a weaving run that saw him beat four players. A second try was disallowed to the home team shortly afterwards because the ball had touched one of the spectators. Whiteside scored the second just before half time to take Auckland out to a 7-0 lead.

During the interval there were no fists thumped into palms, no team talks and no desperate exhortations from coaches. Instead the two teams mixed freely and chatted to each other while enjoying their refreshments.

After the break there were some robust forward exchanges but Auckland got the upper hand with two more tries for an 18-4 victory. But though the rugby was finished the tour was not over yet. That night New South Wales attended a function given by the Mayor of Auckland. The following evening they were guests at the Opera House. The next day the first overseas rugby team to tour New Zealand were farewelled from Auckland Harbour on the start of their journey home.

3

TARANAKI "FAMILIES" DEFEAT BRITISH EMPIRE

16 May 1888, The Racecourse, New Plymouth

Last minute try disallowed by the brother ref

The Great Britain team that toured New Zealand in 1888 was neither an offical one nor truly representative of British rugby. The tour was privately promoted by two international cricketers, Arthur Shaw and Arthur Shrewsbury. The team mostly consisted of men from the northern counties of England and the Border counties of Scotland. The Welsh and Irish were not invited. One player, J.P. Clowes, was banned a few days into the tour when he admitted to accepting 15 pounds from the team's manager Mr Turner to buy clothes for the trip. He was declared a professional, a more heinous crime than murder in those days.

At that time Taranaki was not even a union (the union was officially formed in 1889) but a collection of clubs. In fact the match was very much a family atmosphere with Taranaki being captained by Alf Bayly. Alongside him in the team his brother Charles was playing at halfback. For good measure yet another brother Frank was the referee. Also in the Taranaki team that day were another set of brothers, Alan and Harry Good.

Before the match against Taranaki, the Great Britain team had taken the scalps of Otago and Canterbury and would later defeat Auckland, so what chance did the country boys have when they were not even a proper union?

About 3,000 people came to find out on a beautifully sunny day. Straightaway it looked as if the British team would have the measure of their rivals as their forwards began to dominate. On several occasions their three-quarter Anderton almost got through brilliantly to score, but desperate defensive work by Kivell and Tate of Taranaki kept him at bay.

It was the family connection that scored for Taranaki. Alf Bayly made a dashing run that almost took him over the British line. The ball was fed back to Harry Good who crashed over for the try to give Taranaki Clubs a 1-0 lead. [At that time scoring values were one point for a try, another 2 if you converted it and 3 points for any other form of goal.] Though the try was scored close to the posts, Coghill failed to convert.

The British team, unbeaten at this point on their tour, threw everything into the attack to try and save the match. Just as desperately Taranaki Clubs tackled anything that moved to try and retain their lead. Then the British player Harry Speakman who was a drop goal specialist had a shot at goal. The ball only just missed.

Still just one point divided the two teams as the end of the match approached. With just one minute of play remaining, Great Britain made one final assault as Taranaki Clubs put up desperate defence. There was a scramble near the corner and a British player dived over. But the referee and the Taranaki umpire both conferred and agreed. No try.

When the whistle blew the Taranaki Clubs had achieved a momentous victory. The British team sailed into Onehunga for their next match against Auckland with the black and amber Taranaki colours fluttering from the masthead of their ship.

4

THE GENTLEMANLY NATIVES

16 Feb 1889. Blackheath, England

Rugby, knickers and vulgar gazes

The 1888-89 Native team played an astounding total of 107 matches on their 1888-89 tour of Britain, Australia and New Zealand. They won 78 of them, drew six and lost 23 (the records are not exact but these statistics are generally accepted as close to being accurate as is possible).

The founding father of the Native team was the great Maori player Joe Warbrick who in February 1888 had planned to form a team of Maori or part-Maori players to play against a touring British Football team. That match never eventuated but from that spark Warbrick instead conceived taking a Maori team to tour Britain. Englishman Thomas Eyton had the same idea and the two joined forces with Warbrick acting as captain of the team and Eyton putting up the necessary capital and acting as treasurer. A publican, James Scott from Gisborne, was brought in as manager.

Joe Warbrick selected 26 players for the overseas section of the tour including 5 Warbricks, himself and his brothers William, Fred, Arthur and Alfred. Also chosen was Tom Ellison (also known as Tamati Erihana) who was later to become a great captain, player and innovator. The name of the team had to be changed from "Maori" to "Native" when five Pakeha players were included in the selection, though two of them were not even native. Mac McCausland was born in Australia and Pat Keogh in Birmingham, England.

The Native team started its tour in New Zealand where it defeated Hawke's Bay 5-0 on 23 June 1888. The team played eight more matches in New Zealand before crossing the Tasman and playing two matches. A long boat cruise through the Suez Canal took them to England. Their first match was played against Surrey at Richmond on 3 October 1888. Though the tour by its very nature and number of matches was an epic, the most memorable match was when the Natives (often billed in England as the Maoris) took on the might of England at Blackheath in February 1889.

"By a quick wriggle, however, he escaped but left a portion of his knickers in my possession..."

It was said that England only agreed to play against the Native team because none of the other three home unions would play against England because they had refused to join the International Rugby Board which the other unions had created.

Perhaps the Native team should have borne this in mind when they allowed the Chairman of the English Rugby Football Union, Rowland Hill, to referee the match.

He allowed two English tries where the Native team claimed to have touched the ball dead in goal first on both occasions. The most notorious incident however was the third try when Tom Ellison attempted to tackle Englishman Andrew Stoddart. In Ellison's own words, "By a quick wriggle, however, he escaped but left a portion of his knickers in my possession. He dashed along and the crowd roared; then suddenly discovering what was the matter he stopped, threw down the ball, and in an instant we had the vulgar gaze shut off by forming a ring around him."

The Natives believed Stoddart had called "dead ball" but while the New Zealand players were protecting Stoddart's modesty, another English player picked up the ball and scored with Hill awarding the try. Three of the Maori players walked off the field in disgust but later returned. Afterwards the Rugby Football Union demanded an apology for this behaviour. Threatened with cancellation of the remainder of the tour Mac McCausland, captain in place of the injured Joe Warbrick, was forced to send a telegram expressing, "regret for their behaviour on that occasion." England won 7-0 and the Natives received little reward for protecting Stoddart's modesty from "vulgar gazes."

5

THE BUTCHER'S MATCH

11 September 1897, Carisbrook, Dunedin

One of the most infamous encounters in New Zealand rugby history

In 1897 as Wellington travelled to Dunedin to "do battle" with Otago, there was already bad blood between the two teams. The match the year before at Athletic Park had been such a torrid affair that during a break in the play, as yet more battered bodies were being attended to, one Otago forward turned to a Wellington player and said in a low voice, "I have a little book and I am putting down in it all the names of the b......s I have met and I have got you on the top."

This time there were mutterings around Dunedin as early as the Thursday before the game when the Wellington players arrived and were put up in the same hotel as the match referee. But ironically it was the fast and skilful Otago backs who decided how the match would be played when they sliced through to score the first try.

The Wellington captain was William McKenzie, better known as Offside Mac, the eldest of five famous New Zealand rugby brothers. He was described as a "very keen, hard player" and was the inventor of the fringing wing forward position.

Seeing the Otago backs start to cut loose, Offside Mac immediately ordered his forwards to tighten up the game. They took him at his word and turned the tussle into trench warfare, thundering up the middle of the ruck, knocking aside any and all obstacles to the ball. The forward rushes and the fierce tackling had three Otago players limping or carried off the field.

During the onslaught Wellington scored a try which was greeted by "almost perfect silence" from the home crowd. But they did not remain quiet for long as the action on the field got even more heated. It was Offside Mac who attracted most of their ire. He virtually shut the right eye of Otago's Tommy Wood with a knee to the head and after a collision with Offside Mac, Otago captain Jimmy Duncan came out of it with his face looking the worse for wear.

Hardcastle was another Wellingtonian who was not very popular. When one of the Otago spectators implied Hardcastle's parents had never married, Hardcastle stormed out of the lineout in search of him. Only very reluctantly was he persuaded to return to the field of play.

"Butchers, slaughtermen," were the least offensive terms the crowd used. The match finished with Wellington winning 10-6 but the fight was not yet over. The crowd surged menacingly onto the field and Wellington fullback Billy Wallace recounted in the *NZ Sportsman* how one woman had dug the iron point of her umbrella into various parts of his anatomy with the words, "You dirty butcher. Eh! Eh! Eh!" with every "Eh" accompanied by another sharp prod. On the way back to the hotel the team was pelted by mud and stones from a crowd of 2,000 who continued to chant, "Butchers, slaughtermen."

With the battle won, the Wellingtonians beat a hasty retreat.

"I have a little book and I am putting down in it all the names of the b......s I have met and I have got you on the top."

6

NEW ZEALAND WINS AT VICTORIAN RULES FOOTBALL

15 August 1903, Sydney Cricket Ground

NZ wins first rugby test match on a cricket field and score points at Aussie Rules during a rugby match

In 1903, for the first time, a full New Zealand team was chosen to make a tour of Australia. It was on that tour that New Zealand played their first full international. The man who had the honour of being the very first New Zealand captain was Jimmy Duncan, the Otago captain, who was a brilliant tactician and is credited with formulating the five-eighth system of back alignment. Dave Gallaher was playing at wing forward in that first match but could also lay a claim to being the first New Zealand captain. When fighting in the Boer War, he had captained a New Zealand Army side which had defeated a South African Army side.

Also in that first New Zealand team to take on Australia were "Opai" Asher, famed for hurdling would-be tacklers and the great Billy Wallace at fullback. Wallace kicked the first goal of the game after Evans of Australia was caught offside at a ruck. Australia levelled the score with a penalty. Those were the last points the Aussies scored as New Zealand ran in three tries. Wallace kicked a conversion, a penalty goal and two goals from a mark (then worth 4 points) in the 22-3 victory.

However, earlier in the tour, Wallace had taken an even more important kick. It all began when the New Zealand team found they were sharing a hotel with a couple of Victorian Rules football teams who were staging an exhibition match in Sydney to try and introduce the sport to the city.

The players of each code soon got around to comparing their relative kicking abilities. The Victorians claimed they were better kickers of the ball but Jimmy Duncan told them, "Come to our match tomorrow, we have a man who will show you."

Sure enough the Victorians turned up and during the match, the New Zealand halfback "Skinny" Humphries called a mark on the half-way line. At that time the player taking the mark was allowed to elect another play to take the kick.

Duncan told Humphries to nominate Billy Wallace. Wallace shook his head. It was too far out. But Duncan was not concerned about getting points on the board, he was more intent on showing the Victorians in the stand how rugby players could kick.

"There was no escape," said Wallace later. "So I came back about 11 yards behind the halfway line and placed the ball. The ball sailed straight and true, clean between the posts, at a great height, right over the dead ball line, over the heads of the crowd and landed behind them. It was the finest goal I ever kicked. The crowd gasped and I think that the Victorians were satisfied."

7

1904 FIRST SHIELD MATCH

6 August 1904, Alexandra Park, Auckland

First Shield season. Played 0. Lost 0. Shield retained

The Governor of the Dominion of New Zealand in 1901 was Uchter John Mark the impressively titled fifth Earl of Ranfurly. That year he decided to donate a cup for provincial rugby competition in New Zealand.

When it arrived it was found he had upgraded the original idea to a shield which had on it 12 smaller shields to record the names of the holders (more shields were added from 1935 onwards as more new unions won the trophy). Over a period of time, because of it's wooden base, the Shield became affectionately known as the "log o' wood."

There was some initial embarrassment however when it was discovered that the central motif depicted soccer players. Alterations were made to extend the goals higher to form rugby posts and to change the round ball into an oval. Today, only a close inspection reveals that the players are still wearing shin pads and though the ball is in the air, no one appears intent on catching it.

The Governor formally presented the Shield at halftime of the North Island v South Island match in September 1902. The New Zealand Rugby Football Union had decided that the Shield should be a challenge trophy. In order for it to be challenged for, someone had to possess it, so they presented it to Auckland who had been undefeated for the last 6 years.

At the time the Shield had not acquired the almost mystical qualities that it was later to possess and the Auckland representatives said, "Thank you very much," and tucked it away in a dusty cupboard. The Shield was not played for the following season because the rules said the match had to be at home and that year Auckland made a long tour and played all its matches away.

Wellington had the honour of being the very first Shield challengers in 1904. About 9,000 spectators turned up at Alexandra Park in Auckland to watch and a record £505 was taken at the gate. An Auckland side which included Opai Asher and Dave Gallaher was expected to triumph over a Wellington side which included Billy Wallace and Duncan McGregor. But from the start the match was close with the forwards fighting it out. The backs saw little of the ball until just before half time when the Wellington backline started moving the ball in its own half. The ball was sent out to McGregor who crashed through a last ditch tackle to score.

Auckland fought back in the second half and managed to get a try after a charge around the blindside of a scrum on the Wellington line. With ten minutes left, a cross-field kick by Ward of Auckland was gathered by Wellington and ran back at them. A kick took the ball past the fullback Asher and Gray of Wellington won the race to touch down.

The final score was 6-3 to Wellington. The Shield had been lost and won at the first attempt. But the Earl of Ranfurly was not there to hand it over. He had already returned to Britain unaware that he had started a tradition which would grow to become the Holy Grail of New Zealand provincial rugby.

8

A NATION COMES OF AGE

16 September 1905, Exeter, Devon

1905 was the most important year in New Zealand history

In 1905 New Zealand was an underpopulated backblock of the mighty British Empire. As far as the British were concerned, south east of London was Kent. A little further was France and a lot further, right at the end of the world, was a town called New Zealand.

Many New Zealanders still thought of Britain as "home" and it was still largely the case that machinery, culture and even sport in New Zealand had all originated from Britain.

Rugby, it is claimed, was invented by an Englishman called William Webb Ellis in 1823 (even though he was technically cheating at the time). In 1888 an unofficial British team had come to New Zealand and showed the colonials how to play the game, losing only two matches out of 19, beating the best of New Zealand's sides. In 1905, what could this team of farmers, farriers and blacksmiths hope to achieve against the mother country?

That opinion was echoed in England. "What! New Zealand coming to play against the clubs and counties of England? Well, was there ever such a cheek?" one Englishman commented when he found out about the proposal. No one dared to tell him that they were also scheduled to play against the national teams of England, Ireland, Scotland and Wales.

The New Zealand players who sailed out of Wellington on 30 July 1905 on the *Rimutaka* could not have known what to expect. The team that took the field for the first match against Devon wore black jerseys with a leather insert across the chest and shoulders that could be laced tight. On their chests was a silver fern. Against Devon, Billy Wallace also wore a hat, perhaps reinforcing the impression that the New Zealanders were just a team of country bumpkins.

But both the captain of the New Zealand team, Dave Gallaher, and the vice-captain, Billy Stead, were astute thinkers. Jimmy Hunter was a slippery eel whose number of tries on the tour has never been beaten. Then there was the forward Charlie Seeling, a ferocious tackler who often dived several feet through the air before he hit his target.

But Devon was the English County champion and the match was not expected to be close. Even so the sub editor of a London paper tutted and shook his head as he read the obvious transmission error. His paper was published showing the score of the match to read New Zealand 5 Devon 4 – a narrow and unexpected victory to the tourists.

But the real score was much more incredible. The New Zealanders had not just beaten the Devonians, they had slaughtered them, running riot with 12 tries and winning by 55 points to 4. It was this score, the *correct* score which winged its way along the telegraphic lines to a stunned New Zealand.

It was the day a nation came of age. New Zealand, its rugby and indeed its whole attitude to sport would never be the same again.

9

THE TRY THAT NEVER WAS

16 December 1905. Cardiff Arms Park, Wales

NEW ZEALAND HERALD

Dave Gallaher who captained the All Blacks through the triumphant tour of the British Isles, but who missed the crucial Welsh match through injury.

Welsh wizard makes the tryline disappear

The 1905 All Black "Originals" had an incredible record – 34 matches played, 33 won, a colossal 830 points scored and a mere 39 conceded. But bare statistics alone do not create legend. There must be drama and that most elusive element of all – myth. In Wales, that cathedral of passion, they found it.

The All Blacks were carrying injuries and tiring when they arrived in Cardiff to be greeted by an enormous crowd. At the time the Welsh were in the middle of one of their golden ages, their menfolk were hardened by working in pit and at furnace, and it has been said, they had been watching the All Black's progress with steely calculating eyes.

Forty thousand attended the match with many more locked out. Tour Manager George Dixon described the scene with "the New Zealand team chanting a 'haka' and the Welsh team responded by singing "Land of My Fathers" the whole vast swaying crowd, men and women alike, joining in. It was a wonderful sight and the air was electric with excitement."

The dried black ink on the pages of the record books state that Wales scored the only try, winning 3-0. But every New Zealander knows differently.

In the second half, All Black Billy Wallace made a strong run and broke through the Welsh defence. Confronted by the Welsh fullback Winfield, he threw out a pass to Bob Deans who caught it and raced for the line.

In Wallace's own words. "Instead of going straight ahead, he veered in towards the goal line. Teddy Morgan, the Welsh wing three-quarter, was coming across fast from the other wing and Bob was becoming a little exhausted. Bob saw Morgan in time and altered his course to straight ahead and just grounded the ball six inches over the line."

"[Deans] just grounded the ball six inches over the line."

Billy Wallace

The referee John Dallas who, according to George Dixon, was dressed in "ordinary walking boots, no buttons or bars, and clad in ordinary clothing, including the orthodox high collar" was thirty yards behind the play.

Deans left the ball where he had scored and the Welsh halfback Dicky Owen picked it up and placed it back in the field of play. That was where the referee found it when he finally arrived and despite the protests of the All Blacks he failed to award the try.

So the All Blacks lost their unbeaten record.

Over the years an enormous body of "legal" evidence has been amassed around the incident, including "confessions" by Welsh players and the referee decades after the match. Just two years later Bob Deans lay dying. The words he uttered on his deathbed were, "I did score the try."

10

THE ANGLO-WELSH

25 July 1908. Alexandra Park, Auckland

Fun and games for the Anglo-Welsh – except on the paddock

The Irish and the Scottish unions didn't want to have anything to do with a tour to New Zealand in 1908 so the team that eventually toured was made up of Englishmen and Welshmen, and it became known as the Anglo-Welsh. The best of the backs was Johnny Jackett who had played for England against the 1905 All Blacks. He was also a boxer and when the Anglo-Welsh were in Invercargill he officiated as a judge at a boxing match there.

One of the forwards Fred Jackson met and fell in love with a Maori woman on the tour. Their bliss was interrupted however when he was sent home before the second test for alleged professionalism. He later returned to New Zealand to marry his sweetheart and their son Everard grew up to become a prop forward who played six tests for New Zealand between 1936 and 1938.

By the last game of the tour, which was the third test, the Anglo-Welsh had played 16 matches and won nine. They had been hammered 32-5 in the first test at Dunedin, but had managed to draw the second test 3-3 against a weaker All Black team.

They needed to win the third test to square the series, but in the week leading up to the game there were many distractions. After they had lost to Auckland 11-0 on the Saturday they travelled to Rotorua to take in the thermal region and were given a guided tour by Alf Warbrick of the 1888/9 Native team and the famous guide Maggie Papakura. There followed several days of tours and late night balls which probably put the tourists in the wrong frame of mind for a rigorous rugby

"Man overboard!" went the cry.

match. Though in those long gone days it was likely the locals were just being hospitable rather than having a deliberate plan to throw the Anglo-Welsh off their guard.

Either way they were well beaten in the third test. If New Zealand had kicked accurately the score would have gone well over 40 instead of the 29-0 drubbing. That meant that in the last three games of the tour the Anglo-Welsh conceded a total of 45 points and scored exactly none.

Wing forward George Gillett and the strongly built All Black forward "Bolla" Francis scored two of the nine All Black tries that day. As each dived over the line, neither man could have imagined that they would soon be repeating the same action in quite difference circumstances.

The Anglo-Welsh were seen off from Auckland wharf by a crowd of several thousand including some All Blacks. One of the Anglo-Welsh, Percy Down, was leaning over the rail of the ship to shake hands when he suddenly lost his balance and pitched into the water.

"Man overboard!" went the cry. Down was dressed in a heavy overcoat and struggled to keep afloat in the water. Immediately Bolla Francis and Johnny Jackett dived into the water. They were joined by George Gillett. It was Gillett and Francis who held Down afloat until a rope could be thrown to him and he was dragged out of the water. Apart from the embarrassment and getting soaked, Down was none the worse for his little adventure.

11

THE FEUD BEGINS

10 September 1910. Alexandra Park, Auckland

The first epic match of the Auckland-Canterbury Ranfurly Shield rivalry

There have been many great matches in the history of the Ranfurly Shield and many great teams, but the greatest teams and the greatest rivals have undoubtedly been Auckland and Canterbury, the colossus of the North Island against the colossus of the South Island. Between them they account for almost half of all Shield matches won and when they meet the rivalry is always incandescent. The first real cliffhanger, in what was to become a long saga, took place in 1910.

After Wellington won the Shield at the first attempt, they took it back to the capital. There two weeks later, the first official challengers to a team that had actually won the Shield on the paddock were Canterbury. But they were repulsed. In 1905, though short of seven All Black "Originals", Auckland went down to Wellington and took the Shield back again.

The following season on 11 August 1906, Canterbury was again the first challenger to the new holders. They came up to Alexandra Park and took the strong wind in the first half but were literally blown away by Auckland 29-6 by the end of the match.

It was another four years before the two again contested the Ranfurly Shield and this time the intense North/South rivalry began in earnest. Both teams went at it hammer and tongs before 8,000 spectators but both remained scoreless until just half a minute before half time when Doell dropkicked Canterbury into a 4-0 lead.

In the second half Auckland struggled to get back into the match amidst mutterings in the crowd that it might be good for rugby if the Shield were to spend some time elsewhere. Then Auckland went on the attack from the centre of the field. Hayward broke away and made a great run, passing on to the winger Harrison who beat the tackles of four Canterbury defenders to score alongside the posts.

It was now 4-3 to Canterbury and with the conversion Auckland would move ahead. But there was a gasp of dismay from the crowd when O'Leary missed the easy kick.

But there was a gasp of dismay from the crowd when he missed the easy kick.

From the restart the Auckland forwards went onto the offensive, bursting through the tackles of the opposition forwards. The ball was worked to Harrison who once more went over the Canterbury line. The crowd rushed onto the field in relief, but it was short-lived. The referee ordered the spectators off the field. He had called play back for a scrum.

In the dying moments of the game Auckland launched a desperate attack which saw Francis held up just short of the line. The crowd shouted with relief as Hall dived over to score.

But again the referee disallowed the try and called play back. Was fate against an Auckland victory? Was the Shield destined to cross Cook Strait with red and black colours attached to it?

With just two minutes left to play, the Auckland pack got hold of the ball and drove for the line en masse. The Cantabrians tried to hold them off, but this time the referee blew for the try when Wylie touched down. Auckland had squeezed home 6-4.

12

THE FIRST GREAT SHIELD ERA

16 August 1913, Auckland

Auckland dominant under coaching of Gallaher

Everybody marvelled at Auckland's great Shield era of the 80s and 90s when the province held the log of wood for eight years. "You can look at it, but you can't touch it," was the catchcry as a succession of visiting teams were repulsed. Yet Auckland had already held the Shield for the same period of time during the early part of this century, though in those days they did not play quite so many challenges (just one in 1909).

The sole selector for that period and the architect of their domination was the ex-Originals captain Dave Gallaher, who was undoubtedly one of the great pioneers of rugby in New Zealand. There were some close calls during that eight-year period. Wellington pushed them to a draw before 15,000 spectators in 1910 and Otago did the same in 1912. Marlborough had no such luck in 1911 when they were forced to telegraph the Auckland Union cancelling their first ever challenge because they were not able to muster a strong enough team (they had been beaten by Auckland 32-0 in 1908 in a non-Shield fixture).

In 1913 Taranaki came to town to try and take the Shield home. They had already made a valiant effort the year before but lost by just one point. This was the last match of the season, so it was the last chance anyone would have to give the Shield a summer holiday outside of Auckland.

The sole selector for that period and architect of their domination was the ex-Originals captain Dave Gallaher.

A crowd of 8,000 came to watch Auckland's 24th defence on a sunny day with a slight breeze across the ground. The local supporters were concerned to see Taranaki's forwards take early control. Within five minutes of the start Taranaki had moved the ball out to Roberts who kicked ahead over the Auckland line and was first in the race to touch down.

Auckland struck back just before half time. The ball went loose, Macky pounced on it and spread it out to Geddes who went over for the try. When O' Leary converted, Auckland went into the lead 5-3. But Taranaki came straight back after the break. Pressure on the Auckland full back O' Leary forced him to make a weak kick which was gathered up by Taranaki to score. A further try to Dewar from a scrum took the lead to 9-5.

After eight years of holding the Shield Auckland were not yet ready to hand it over so easily. When they got the ball out to Duncan McGregor on the wing he took off for the corner. Meuli the fullback raced across to tackle him. They crashed across the line together with McGregor scoring.

Just one point behind, Auckland ran the ball again from their own 25 yard line, rapidly moving the ball up the field through Geddes to Weston. He had only the fullback to beat with two players unmarked outside him. Unexpectedly he cut back and scored himself. At 11-9 it looked as if Auckland had done enough to retain the Shield.

But Taranaki continued to drive forward. Just a few minutes before time the ball was won at a lineout and moved out to Hill on the wing. He ran for the line, but was blocked just short. He passed to Dewar who was following up. The Auckland defence was stretched and when Dewar passed to Roberts there was no one left to tackle him. Even without the conversion, which Hawkins kicked anyway, the Shield was heading south and the first great Shield era was brought to an end.

13

HOT DRINKS IN THE GALE

6 September 1913. Athletic Park, Wellington.

Aussies drowned out of the game in Wellington

Before the First World War rugby still tended to be less structured than it is today. When the All Blacks played against Australia at Athletic Park in 1913 the Wellington wind was up to gale force and so the decision was made, for the benefit of the players, to divide the match into four quarters.

That day *The Dominion* reported, "Only about 5,000 people forsook the fireside or other haven of cosiness for the bleak, clay-banked paddock which we call 'the Park'."

The reporter went on to imply that the rain and wind was caused by the "weather imps" which he claimed were as numerous in Wellington as were the thieves in the story of Ali Baba. He thanked the local union for the vantage point he had been afforded for the match from which the press could, "ascertain with certainty how strong the wind was, how wet the rain was and what excellent shower bath was obtainable by anyone who momentarily took one eye off the

"...what excellent shower bath was obtainable by anyone who momentarily took one eye off the overflowing roof gutter."

overflowing roof gutter."

There was even less shelter for the players on the field and some of the Aussies were tempted to snatch the umbrellas out of the hands of the crowd. They were more used to the sun beating down on their heads, playing on dust instead of in mud.

With the wind buffeting their backs in the first half, the Australians made a fight of it, but even when they were reduced to 14 men through the loss of the halfback Taylor, the All Blacks were too good for the visitors.

As the match went on, the rain became worse and by the third quarter the ball was like a slippery cake of soap. It was better to kick it or keep it in hand and run with it. That's exactly what Gray the New Zealand first five-eighth did on an amazing run during which he sidestepped virtually the entire Australian team to score under the bar.

It was all too much for the shivering Aussies as the All Blacks scored a total of eight tries, three of them to the South Canterbury winger William "Tiger" Lynch. McKenzie scored two and the others went to Murray, Gray and Roberts, with Roberts also kicking three conversions in the final score of 30-5.

At that point the Australians would have been happy to return to the sunshine of their homeland. However it was the All Blacks who travelled, to the sunny climes of California, leaving behind a second-string All Black side to polish off the Aussies in the next two tests (Australia won the third when the sun finally came out).

That first test is believed to be the only international match to be broken into four quarters instead of the normal halves and was certainly the only one where the players were served hot drinks during those breaks. Some of the players even took the opportunity to change into dry clothing.

14

NEW ZEALAND BEAT ALL-AMERICA

15 November 1913. Berkley, California

All Blacks toll the bell for the death of rugby in the USA

The reason why rugby has never taken a grip on the USA is simple – the Americans like to win at everything. This is best exemplified in the America's Cup Yacht races where the rules have been bent and even re-invented to suit the circumstances and ensure the Americans retain the "auld mug."

Cricket was a sport the Americans simply could not understand; so instead they invented baseball. Rugby is undoubtedly a sport they could have become champions at. They have men of size and pace to tangle on the world stage with the likes of the All Blacks and South Africa.

When the All Blacks toured in 1913, rugby was the major form of football played in California. The big games there often drew huge crowds. New Zealand played an All-America team that was chosen from Californian Universities and clubs, but the game is officially recognised as a test match.

Among the All American team was an Australian called Danny Carroll who was something of a rugby gypsy. Not only had he represented his home country Australia in 1908-9, but in 1908 he had also played for the English county Cornwall who were representing Britain at the 1908 Olympics. Ironically they were defeated by Australia 32-3.

About 10,000 people sat on the "bleachers" to watch the match between All-America and the All Blacks. Sawdust had been scattered on the field to try and damp down recent rain. That firmed the ground sufficiently for the spectators to watch a fast and open game which New Zealand won convincingly by 13 tries to nil. The third try was scored by McKenzie in the middle of a mud pool, no doubt bringing back memories of the match in Wellington a few weeks before. He changed into a clean jersey as the conversion was being kicked. The final score was 51-3.

The All Blacks were dominating on that tour. They scored a massive total of 610 points, conceding just 6, an incredible ratio of 102 points scored for every 1 conceded.

They scored a massive total of 610 points, conceding just 6, an incredible ratio of 102 points scored for every 1 conceded.

Though the USA are still the current Olympic Rugby champions (they won the title in 1924, the last time rugby was played at the Olympics) the game eventually perished as a major sport there. Instead the Americans went on to invent their own form of "football". A scrum became a scrimmage, the players got bigger, the field became narrower, the playing time shrunk to one hour, except it was really four hours long so they could fit in hundreds of commercial breaks.

R.I.P. the USA as a major rugby force.

15

SOUTHLAND "BORROW" THE SHIELD

15 September 1920. Rugby Park, Invercargill

A South Island team wins the Shield for the first time

For the first 18 years of its existence the Ranfurly Shield was the sole property of North Island teams. Southerners had cast a greedy eye on it from the very beginning, Canterbury and Otago both travelling north to become the second and third challengers (after Wellington won it at the first attempt).

By 1919 there had been only three holders – Auckland, Taranaki and Wellington. That year Wellington had the Shield firmly in its grasp again and decided to take it for a little jaunt across Cook Strait. Canterbury gazed at it with envy, but had to wave goodbye to it when Wellington easily repulsed their challenge 23-9.

The following year the Wellingtonians decided to flaunt their supremacy some more, taking the Shield with them for the annual fixture against Taranaki at the Hawera Show. They even had the cheek to show it off at the Auckland Domain and were lucky to get it back again after winning a seesaw match against Auckland by the narrow margin of 23-20.

But the confidence of the Wellingtonians knew no bounds and they decided it was time the Shield saw some more of the South Island. South Canterbury were despatched 32-6, Otago were felled 16-5 and the Ranfurly Shield roadshow rolled on into deepest Southland.

At the time Southland held the unenviable record of being the most unsuccessful challenger for the Shield when in 1906 they were thumped a record 48-12 by Auckland. But this was Wellington's eleventh Shield match of the season, the fifth away from home and it was also a mid-week fixture. Were they getting tired and overconfident?

...this was Wellington's eleventh Shield match of the season, the fifth away from home...

The Southlanders took the brisk wind and their forwards dominated from the start. They went ahead through a penalty goal by "Son" White. Soon after the ball was booted ahead and the race to it won by Fortune. 6-0. Before the Wellington team could fight back the Southland forwards again went on the attack and heeled the ball to their backs. Billy Stead ran with it, beat the defence and was able to touch down under the crossbar.

At 9-0 down in the second half Wellington tried to come back at the challengers but Southland pulled further away with a try to Holmes. Finally Wellington managed to get on the board with a try to McCabe. That made it 12-3. Then Mark Nichols kicked a penalty, 12-6. Was this the start of a late resurgence from the holders?

No. Stead scored again and the conversion made the final score 17-6. For the first time a South Island province had got their hands on the Shield. Wellington had finally lost the Shield after resisting 15 challenges (10 of them in 1920) and they had started the tradition of taking the Shield on tour with them.

The following season Southland returned the favour and took the Shield on tour up to the capital. Wellington promptly snatched it back again.

16

THE OLD AND THE NEW

13 August 1921, Carisbrook, Dunedin

The historic first test match against the Springboks also saw the recall of the oldest player ever to appear for New Zealand

South Africa was a late starter at rugby. A British Isles team had toured the country in 1891 and conceded just one try in winning all their 19 games. But by 1912 the Springboks had learned quickly and sent a team to the British Isles that was good enough to win all its test matches. It was inevitable that the two southernmost outposts of the former British Empire would finally meet on a rugby field. A tour was arranged but had to be postponed when the First World War intervened. On 12 July 1921 the first touring party of 29 Springboks arrived in Auckland aboard the *Aeneas* and received a warm welcome.

The Springbok team had played nine games as they came into the first test at Dunedin. All had been won except for a draw with Taranaki and a 6-4 loss to Canterbury.

For this match the All Black selectors recalled Ned Hughes who had made his debut for New Zealand in 1907 and last played for them in 1908. His All Black career had been interrupted not just by the war, but also by becoming a league player, at which he also represented New Zealand.

It was not clear if the selectors were aware of his real age at the time. He claimed to be a sprightly 36. Some said he was actually 42.

Between his last All Black appearance and this one there was a gap of 13 years, two months and one week. It was not clear if the selectors were aware of his real age at the time. He claimed to be a sprightly 36. Some said he was actually 42. His army records show he was born on 26 April 1881 which would have made him 40 years old. Hughes was one half of the New Zealand front row which in those days packed down as a two-man unit.

The South Africans were the first to score when van Heerden went over in the corner following a scrum. The Springboks almost went further ahead into the second half, but then the All Blacks came back. A high kick evaded the South African fullback and Belliss got the ball to score.

The All Black's next try is considered to be one of the classic test match tries of all time. A scrum was set in midfield and the ball was hooked back on the All Black side. Halfback Ginger Nicholls fed first five-eighths Ces Badeley who punted towards the right wing of West Coast speedster Jack Steel. Steel took the ball awkwardly and clutching it to his shoulder blade, set of on a blistering run still fighting to gain control of the ball. He beat fullback Morkel and centre Meyer on his 50-yard sprint but he only truly gained possession of the ball as he crossed the line. A third try to Storey gave New Zealand a 13-5 win.

When Hughes played in the second test at Eden Park two weeks later (which New Zealand lost 9-5) he was aged 40 years and 123 days. His long All Black career came to an end when he was dropped for the third test which was drawn 0-0 in atrocious conditions at Wellington. The series was tied and the All Black-Springbok rivalry had begun in earnest.

17

THE MOST UNFORTUNATE MATCH EVER PLAYED

7 September 1921. McLean Park, Napier

Rugby overshadowed by the impending spectre of apartheid

New Zealand and South Africa knew very little about each other and their respective countries when they played their first test series against each other in

1921. It would be another 60 years on the 1981 Springbok tour before that difference would become glaringly apparent.

Maoris had been key members of representative teams since the start of the sport in New Zealand, but Maori team rugby started in 1910 with a tour to Australia which saw just two defeats in 11 matches. Thereafter a Maori team was selected to play against the 1913 Queensland tourists (the Maoris won 12-9) and South Africa in 1921.

There was already some ill feeling between the teams prior to the kick off when the Springboks turned their backs on a group of Maori girls performing poi dances and singing. "We were seething with anger as we waited for the kick off," one Maori player said later.

For their part the Springboks disliked the Maori scrum which in the New Zealand fashion had two men in the front row. When they engaged the Maori team turned their scrum into a flying wedge which drove virtually through into the second row of the Springbok scrum.

The South Africans scored first with a try to Zeller, but Tapsell equalised with a penalty. There was some "vigorous" play in the first half with several players requiring medical attention. In the second half the Springboks had the advantage of the wind and kicked deep into Maori territory, taking the lead with a try to van Heerden. But it was a game that the Maoris came very close to winning.

A mark by Jack Blake in front of the Springbok posts (which he would probably have successfully kicked) was disallowed by the referee because, he said, the ball had touched another player before Blake had caught it. The South Africans got the ball off the surprised Blake and raced downfield to score. Some thought that the South African Zeller had put his foot into touch before making the telling pass, but the referee allowed the try. The Maoris were doubly unlucky.

Yet still they were not out of the game and the Maori forwards went on the rampage, continually pinning back their opponents until the whole pack crashed over the line between the posts for a try that was awarded to Garlick. Tureia converted, but the match ended with the Maoris losing by just one point, 8-9.

A notorious incident occurred after the match. A New Zealand telegraph operator made copies of a South African correspondent's report which read:

"Most unfortunate match ever played....Bad enough having play team officially designated NZ natives [actually called NZ Maori XV] but spectacle thousands of Europeans frantically cheering on band of coloured men to defeat members of own race was too much for Springboks, who frankly disgusted."

He showed these around the office and one copy made it's way to a Napier billiard saloon where it was snapped up by a local journalist and published in the Napier *Daily Telegraph* to great furore.

...the Maori forwards went on the rampage, continually pinning back their opponents until the whole pack crashed over the line between the posts for a try...

18

THE SECOND GREAT SHIELD ERA BEGINS

9 August 1922. Athletic Park, Wellington

Hawkes Bay wins the Shield on a Wednesday

It seemed Wellington still had something to learn about retaining the Shield. In 1920 they had taken it on tour and lost it. They got it back again in 1921. The first challenge of the 1922 season was a warm-up at home against the country boys from Hawkes Bay. But the Bay boys had a good team even if Maurice Brownlie was overseas with the All Blacks. They also had one of the most astute coaches ever in Norman McKenzie who was "sole selector" of the team from 1916-1945.

McKenzie was a big man in every way. He knew his own mind and often expressed in it pungent, sharp words. He could be dictator or diplomat, had a good sense of humour, was a gifted raconteur and had a memory "a yard long."

The Hawkes Bay Rugby Union had been very disorganised when he went there from Wairarapa during the First World War. He travelled around the region scouting out a wealth of new talent over the years. As well as the Brownlie brothers Maurice, Cyril and Laurie, he also unearthed a 17-year-old Maori lad who went by the name of George Nepia.

McKenzie shrewdly calculated that his unfancied team might take Wellington by surprise in a midweek match. A crowd of 4,000 turned up to enjoy the good weather and to watch Wellington have an early season romp. But Hawkes Bay went straight onto the attack and after ten minutes Kirwan scored a try. Wellington came back to level the score, but then Hingston made an opening for Hawkes Bay, the ball was moved wide to Scott who touched down. With the conversion and a penalty goal the Bay shot out to an 11-3 lead.

He could be dictator or diplomat, had a good sense of humour, was a gifted raconteur and had a memory "a yard long."

The Wellington team had to wake up from their Wednesday afternoon nap or they would lose the Shield. They came out strongly in the second half and Udy made a weaving run to bring the score back to 11-6. Just when it seemed the holders might claw their way back into the game, the Hawkes Bay forwards surged again and McNab scored in the corner.

As Wellington grew more desperate to snatch back the match and the Shield, the Bay were able to counter attack and their centre Blake managed to dodge several high tackles to score near the posts, making it a simple conversion for Yates.

That took the score to 19-6. Wellington fought to the last with Ryan scoring a try to bring the score back to 19-9, but the country boys had fooled their city cousins and took the Shield.

19

WELSH FOUNDER ON NEPIA ROCK

29 November 1924. St Helen's, Swansea

Hori George Mary Nepia saves the day

Among those selected to go on the 1924 All Black tour to the British Isles was a young 19-year-old Maori from the East Coast called George Nepia. He was chosen at fullback even though he had only ever played in that position at the All Black

PETER BUSH

The Welsh players seemed in awe of George Nepia in full cry during the test at St Helens, Swansea.

"By sheer strength, they barge their way through with the ball and there stands Nepia alone between them and the desired objective..."

trial matches.

After a short tour of Australia, the All Blacks went on to cut a swathe through the British Isles, beating all that was set before them, but the match they really wanted to win was the one against Wales – to avenge the ignominious defeat of the Originals 19 years before.

In glorious weather 50,000 Welshman packed out the St Helen's ground at Swansea hoping they wouldn't. The match had a nervous start with the first four balls offered were rejected by the Welsh captain Jack Wetter. The fifth was found to be acceptable just as the crowd were starting to hoot.

New Zealand scored four tries during the match, but the real triumph was in keeping the Welsh scoreless. In those days dribbling the ball before the forwards was a great art and it took a fullback of great courage to deal with it.

Afterwards the local paper the *Western Mail* wrote, "Nepia...has never given a better display than on Saturday. There were times when it appeared that nothing would stay the fierce rushes of the Welsh pack. By sheer strength, they barge their way through with the ball and there stands Nepia alone between them and the desired objective...His judgement is uncanny and his pluck magnificent."

In his autobiography, *I, George,* Nepia described one such incident. "The Welsh forwards came down the field with the ball at toe. Wetter was in the van, I back-pedalled. It meant loss of ground, but I knew what I was after. I was seeking to drive them to one side and by compelling their attention as I moved I was diverting them from their task of watching the ball. As their eyes shifted from the ball to me, I pounced. I swept in, gathered the ball, shouldered my way through them and got my kick away."

With the emphatic 19-0 victory, the defeat of the Originals was avenged with interest.

On that tour Nepia not only gained accolades as the greatest fullback in the world, but a new name as well. It was the practice then to sign the menu cards at the after match function. When confronted by the eminent name of an opponent, one R.H. Hamilton-Wickes, Nepia decided to extend his own name from plain old G. Nepia to Hori (the Maori for George) George Mary (his mother's name) Nepia. For many years afterwards he signed all requests for autographs as H.G.M. Nepia.

20

OFF!

3 January 1925, Twickenham, England

The blackest day in All Black history

The Invincibles arrived at Twickenham to play England three days into the new year in 1925. The referee for that match was a small Welshman called Albert Freethy. He had already refereed the All Blacks on five previous occasions during the tour.

Prior to one of these games, Cyril Brownlie had been joking to some of the other All Blacks that he had not yet scored a try on the tour. They promised that if the proper moment arose, they would set one up for him. During that match, with the All Blacks well ahead, an overlap materialised on the outside. Cyril Brownlie was positioned inside, so it was completely unexpected when the All Blacks moved the ball back to him to score his try.

The referee Freethy was also taken by surprise and got in the way as Brownlie thundered towards the line. As a result Brownlie failed to score. At the aftermatch function Brownlie jokingly said to Freethy. "Hey, ref I'd have my try if you hadn't been in the way, you silly bugger."

Freethy didn't appreciate the rugged New Zealand sense of humour. But did he remember the incident when the England match started in such a rugged fashion? Three times he issued a general warning to all the forwards. Eight minutes into the contest he blew his whistle again. The ground stilled at his outstretched arm. Slowly the tall figure of Cyril Brownlie made his way across the turf with his head bowed.

The All Black captain Jock Richardson appealed to both the referee and the English captain Wavell Wakefield, but both shook their heads. Wakefield did not protest because in England the referee's decision was considered final.

Freethy claimed Brownlie had kicked another player (Brownlie later denied it). Brownlie walked on towards the tunnel watched by the silent crowd including HRH Prince Edward the Prince of Wales in the royal box. "Never in my life have I experienced anything like the weird silence that fell over Twickenham as he walked away," All Black Read Masters later wrote in a book on the tour.

Slowly the tall figure of Cyril Brownlie made his way across the turf with his head bowed.

When the other All Blacks came into the changing room at half time they found Cyril Brownlie in tears. The Prince of Wales came down from the royal box and tried to remonstrate with officials to have Cyril Brownlie reinstated, but to no

NEW ZEALAND HERALD

The English players seem bewildered, the referee's arm points and Cyril Brownlie's head is down as he becomes the first man to be sent off the field in an international match.

avail.

On the field the All Blacks were reduced to 14 men (13 if you consider that wing Jack Steel was injured) but they also had a Brownlie in Maurice who played like two. He scored a try when he forced himself over the English line from a maul, one of three tries that saw New Zealand win 17-11.

But it was not the great victory that the English newspapers trumpeted. It was the long and lonely walk by the first man to be sent off in an international.

21

LEAVE MR BROWNLIE ALONE

8 August 1925. McLean Park, Napier

The great Maurice Brownlie was the colossus of the 1930s

Great eras come about because of great teams and great teams possess great leadership. While off the field the great Hawkes Bay Shield team of the 1920's had Norman McKenzie, on it they had the imposing figure of Maurice Brownlie.

Brownlie is probably one of the three greatest forwards ever to have represented New Zealand, standing with his muscular arms folded on a rocky plateau head and shoulders above the others along with Colin Meads and Sean Fitzpatrick. Such was Brownlie's stature that legends grew around him. It seems he was the first New Zealand rugby player to be attributed with the folk tale of training by running up hillsides with a sheep tucked under each arm.

The story also goes that when the family car inadvertently went into a river Maurice pulled it out with his bare hands. Whether or not these stories are true there is no doubt that he was the archetypal All Black forward of his era – a farmer whose strength was born out of many hours of physical toil.

Norman McKenzie said of him, "He was the greatest player I ever saw as a side row forward because he was great in every aspect of forward play, but particularly his ability to lead his forwards out of a corner, yard by yard up the touchline."

In August 1925 while still in the midst of their great Shield era, Hawkes Bay was challenged at McLean Park by Southland. It was a wet and windy day and Southland took the advantage of the strong wind to go out to a 12-8 lead at half time. But with the wind behind them in the second half Hawkes Bay pulled away to score 23 unanswered points and win 31-12.

The match has gone down in history for a reported incident where a 19-year-old Southland player, Hazlett, was warned against illegal play by the referee with the words, "Now then, Hazlett, you leave *Mister* Brownlie alone!"

It showed the awe and respect in which even a referee held Brownlie, though it is doubtful if Brownlie needed any help from the referee to protect himself. The following poem was published shortly after the incident.

CROWN STUDIOS

The commanding figure of Maurice Brownlie. Even referees called him "Mister".

A Southland Saga

By Acey

The Southerners all stood a-trembling with fear
As the referee strode up the ground,
In Richardson's eye stood a glittering tear
His emotion, though deep and profound,
Could find no expression in gesture or word-
Could his team such an error atone?
And this was the sentence the Southerners heard:
"You leave *Mister* Brownlie alone!"

22

THE NEW ZEALAND STORM IS ABOUT TO BREAK

19 September 1925. Eden Park, Auckland

The Invincibles on home soil were simply...invincible

In 1925 when New Zealand played against New South Wales the match was then considered to be a "test." New South Wales was effectively Australia as Queensland had recently gone over to rugby league.

The tour that New South Wales agreed to that year was a tough one with 11 matches within a month, travelling by sea, rail and road. They were tiring when they came to the "test" in Auckland. Before the match the All Blacks performed a haka led by George Nepia. It was not the modern "ka mate, ka mate" version but one especially written by Judge Acheson of the Native Land Court and Wiremu Rangi. Translated from Maori the All Blacks chanted:

Nepia: Let us prepare ourselves for the fray.
Team: We are ready.
Nepia: The New Zealand Storm is about to break.
Team: The sound of the breaking.

Nepia: The New Zealand storm waxes fiercer.
Team: The height of the storm.
Leader: Now then.
Team: We shall stand as children of the sun.
We shall climb to the heavens in exaltation of spirit.
We shall attain the zenith.
The power! The power!

One can imagine the voices raising to a final crescendo of intensity as they chanted the last line, *the power, the power.*

The team that day contained 13 Invincibles including Nepia, Cooke, Nicholls and Maurice Brownlie. During the match Brownlie came in for some attention from the New South Welshman, Smith, who was known to be free with his fists. But while the flurry was going on New Zealand scored the first try of the game to Mill. A little later when a ruck broke up, Smith was found to be lying on the ground. The referee asked the All Blacks if they knew what had happened, but none of them had seen a thing.

The All Blacks played the match with the same close knit intensity. They had iron forwards (averaging over a stone a man heavier than the visitors) and fast backs. They had experience and vigour, imagination and commitment. Their attacks were penetrating and were supported in numbers.

At first the valiant New South Welshmen survived, but then withered under the onslaught of New Zealand running and passing the ball despite the wind and rain. McGregor scored, then New Zealand attacked straight from the restart and Finlayson scored. Before New South Wales could recover Brownlie scored another try under the posts.

The visitors fought back to bring the score to 16-10 but it only forced the All Blacks up a gear. They ran the ball up the field. The final pass went to Brownlie who crashed over the line with two would-be tacklers clinging to his back. It was Brownlie who put Cooke in at the corner for a try. Then it was Cooke's turn to set up the score. He punted ahead, caught the man who caught the ball and it was Robilliard who ran in to touch down. Nichols finished the matter with a kick which brought the final score to 36-10.

Such was the dominance of the All Black performance that there are those who witnessed this match first-hand who say this was the finest All Black team ever to play in New Zealand.

23

THE BATTLE OF SOLWAY

9 July 1927. Solway Showgrounds, Masterton

One of the fiercest Shield matches had an even more controversial aftermath

In 1926 Hawkes Bay had an all-conquering season. They won all their home matches by huge margins and had the audacity to take the Shield to Canterbury and defend it there. But in the off-season many players defected to different unions, most noticeably Bert Cooke who went to Wairarapa. The holders were stunned when on 3 June 1927 Wairarapa beat them 15-11 to end their five-year tenure of the Shield.

Hawkes Bay immediately demanded a rematch which was granted by Wairarapa. The Bay coach Norman McKenzie took his players into camp for a week before the game. The Wairarapa coach Ted McKenzie (Norman's brother) made his players sleep at the ground on camp beds under the grandstand. The battle lines were drawn.

On the day of the match 3000 travelled from Hawkes Bay to watch. There were 10,000 at the Showgrounds, the vast crowd encroaching onto the field, narrowing the width of it. Spectators took up every vantage point, hanging from trees and sitting on the roof of an implements shed which collapsed under the weight.

Amazingly the referee for the match was Bert McKenzie, yet another brother in this famous rugby family. But there was no family atmosphere on the field. No quarter was asked for or given and so physical and committed were both sets of teams that the first 40 minutes extended to an hour with all the stoppages for injuries.

...the first 40 minutes extended to an hour with all the stoppages for injuries.

According to the *New Zealand Times* newspaper, at one point Wairarapa's All Black Quentin Donald ran through a ruck straight into the arms of Maurice Brownlie. "No blows were struck, but the pair glared at each other," the paper reported. Bert McKenzie decided to send them both off anyway.

McKenzie probably intended to settle matters down, but instead both teams went at each other with renewed vigour. At half time the score was 5-5. In the second half Hawkes Bay moved ahead with a penalty goal, but then Wairarapa 's Cundy made an interception and passed on to Cooke who scored. Wairarapa now led 10-8.

With 20 minutes to go, Hawkes Bay scored after a dribbling rush. Then after a break from Cyril Brownlie, Swain scored. Wairarapa were finished off with a second try to Huxtable. After a ferocious contest Hawkes Bay had won back the Shield with a final score of 21-10.

Except they hadn't.

Stan Dean, the chairman of the NZRFU was at the match. As the teams walked off, he pointed at a Hawkes Bay player and said to a Wairarapa official, "Why don't you protest against that man?"

"That Man," was Wattie Barclay who ironically had only come on when Maurice Brownlie was sent off. An official protest was made and it was discovered that after being out of the area looking for work, Barclay had been back less than the minimum of three weeks required to qualify for Hawkes Bay. The match and the Shield were awarded to Wairarapa. Hawkes Bay appealed, but to no avail. The Shield was lost in most controversial circumstances.

CROWN STUDIOS

The elusive Bert Cooke who helped Wairarapa take the shield from his former province Hawkes Bay.

24

THE SAGA CONTINUES

1 September 1928. Newlands, Cape Town

Once upon a time...

Once upon a time there were two great rugby playing nations, both of which claimed to be the best in the world. The All Black-Springbok battle for rugby supremacy had begun in 1921 when the first series between the two countries was drawn in New Zealand. In 1928 it was the turn of New Zealand to travel to South Africa to continue the contest.

The team was captained by Maurice Brownlie and the vice-captain was first five-eighth Mark Nichols, but was lacking George Nepia (because he was Maori) and Bert Cooke (because of business commitments).

In a four test series the All Blacks were 2-1 down as they approached the final test in Cape Town. They had suffered administrative difficulties, long and energy-sapping train journeys and differences of interpretation of the laws with the referees as well as robust and committed opposition.

...the All Blacks did the haka and the Springboks responded with a Zulu "war dance" which looked more like choir practice.

With the players in low spirits after a loss to Western Province, Maurice

Brownlie gave very clear instructions for training prior to the last test. Nobody was to touch a rugby ball or partake in any exercise other than a morning walk.

On the morning of the match hail and rain fell in Cape Town. After an attempt to mop up the puddles, sawdust was spread on the most slippery parts of the surface.

This match saw the return of Mark Nicholls who had not played in the first three tests. Nicholls was undoubtedly one of the greatest players ever to play for New Zealand. Not only was he a player of great skill and character, but he had honed his abilities by spending many hours apprenticing under probably the greatest player of the 1905 originals, fellow Wellingtonian Billy Wallace.

Before the match the All Blacks did the haka and the Springboks responded with a Zulu "war dance" which looked more like choir practice.

Nicholls got straight into the action, kicking a penalty goal after just three minutes. South Africa took the lead when Lilburne won the race to a kick ahead and Osler converted. Then Nicholls took a drop kick at goal which hit the upright and Robilliard almost scored from the rebound. It was a Nicholls goal that put the All Blacks ahead again. As rain fell, turning the match into a hard forward tussle the All Black forwards broke through and Swain scored. Nicholls' kick fell short, as the ball was now becoming heavy.

The All Blacks went back on the attack from the restart and won a scrum in the South African 25 from which Nicholls kicked the goal that secured victory 13-5 and squared the series, meaning the two old rivals were dead level once more. Who knows what might have happened if Nichols had played in the previous tests. As the *Cape Times* stated afterwards, Nicholls was, "the life and soul of the backline and an inspiration in himself."

25

THE SOPERS – THE FAMILY TEAM

8 September 1928. Rugby Park, Invercargill

Family Affair defeats Southern Club

There are three things that flourish in Southland, rabbits, gorse and the Soper family. The family's roots stem from Cornwall from whence William Davis Soper

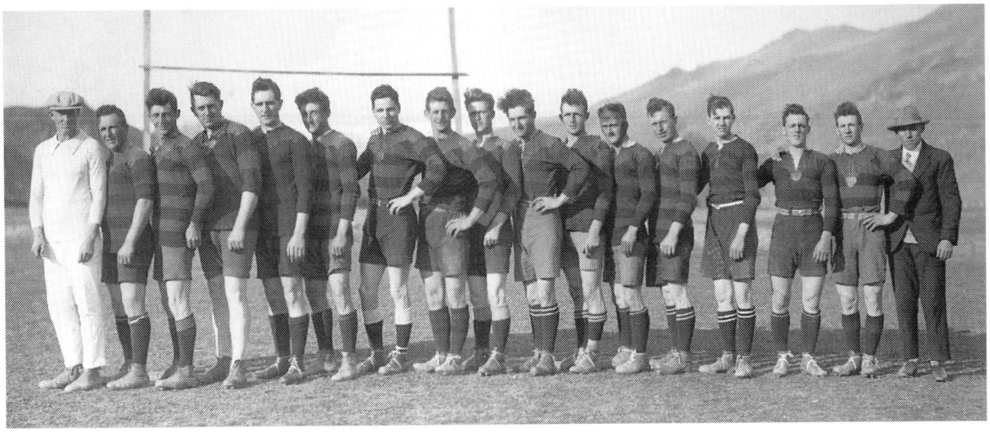
The Sopers.

(generally known as W.D.) emigrated to the goldfields of Australia. There, aged 25, he met and married 14-year-old Hannah Boyer. They later emigrated to New Zealand and produced a total of 16 children, 14 of whom survived to adulthood. Their children were also very productive so that when Hannah Soper (who outlived her husband) died in 1918, aged 76, she left behind 82 grandchildren and 16 great-grandchildren.

By 1925 there were so many able bodied Sopers that, as the old joke goes, they had enough to start their own rugby team. So they did. In 1925 Frank Soper issued a challenge to the local Balfour team. Before the match each Soper was sent a telegram which in the words of Admiral Nelson read, "The Sopers expect every man will do his duty."

The Sopers won, so the next year Balfour sought revenge, but again the Sopers won. In 1927 in what had become an annual fixture Balfour finally won 8-3. The following year after disposing of Balfour 9-0 the Sopers sought a bigger challenge. Southern Club were a strong side and winners of the local Wednesday competition. But on 8 September 1928 they came up against their most formidable opponent yet – the Sopers.

"They can run and leap like deer. As for passin' an' tackling their experience wid dodgin' cows, wild pigs an' motor cars has made them well perfect..."

The match was played before 10,000 spectators as a curtain-raiser to the Southland-Otago clash. There was some concern for the Sopers before the game when one of their key players, big strong Frank Soper, arrived at the train station with just enough time to get to the ground before the kick-off. With "splendid tackling" and a "beauty of a try" when Mark Soper followed up a kick at goal, gathered up the ball as it came off the crossbar and scored under the posts, the Sopers were able to triumph 14-11.

Shortly afterwards a letter was received by the local paper which read, "The Sopers made a name of themselves that'll go down to posterity, an' I shouldn't be surprised to find the team go on tour nixt season, an' lick anything that happens along. It may seem strange, but it's a fact that the Soper boys all know one another, an' have been brought up in a hardy district. They can run and leap like deer. As for passin' an' tackling their experience wid dodgin' cows, wild pigs an' motor cars has made them well perfect..."

26

THE FIRST LIONS

3 June 1930. Athletic Park, Wellington

Wellington records a famous victory

In 1930 there was great excitement when the second official Great Britain team arrived in New Zealand for a 21-match tour (the first had been the 1904 team which played five matches in New Zealand). Unfortunately their captain Wavell Wakefield (later to become Lord Wakefield of Kendal) was forced to withdraw but the team still contained some excellent players including the speedy English wing Aarvold and the almost as speedy Welsh loose forward Ivor Jones.

The tourists arrived at Wellington aboard the *Rangitata* on 17 May where they were welcomed by officials of the NZRFU in a shed on the wharf which had been decked out with flags and greenery especially for the occasion. This was the first British touring team in New Zealand to be called the Lions. According to the book *The Lions* by Wallace Reyburn, the nickname had been given to the British touring team to South Africa in 1924 because of the lion logo on their ties.

...the pressure of those locked outside succeeded in the gates bursting open and letting thousands more into the ground.

The 1930 Lions team played open rugby and contributed to what has been described as "one of the most thrilling games to be played on Athletic Park." In the first four matches of the tour they had polished off Wanganui, Taranaki, Manawhenua and Wairarapa-Bush by large margins.

In the fifth match against Wellington the home province were expected to be strong and accordingly the British put virtually a full test side onto the field.

Every possible vantage point was taken up at Athletic Park long before the game had started and the gates had to be closed. But the pressure of those locked outside succeeded in the gates bursting open and letting thousands more into the ground. Others had climbed onto the roofs of houses surrounding the ground

NEW ZEALAND HERALD

The first British Lions team to visit New Zealand. Here, action from the first test in Dunedin, played in rain and snow.

trying to get a decent view.

Wellington played with the wind in the first half but it was the British who took the lead first – with a try to the Welsh wing Morley which was converted by Black. But the Wellington forwards came straight back at the British and dominated them for the remainder of the half. Porter dropped a goal, then with the help of the wind, Hazelwood was able to kick a goal from ten yards inside the half-way line. McPherson scored a try after good passing between the Wellington backs that gave the home team a half-time lead of 12-5.

Aided by the wind and a superior set of backs the British whipped up a storm in the second half and Wellington was forced to defend desperately. Some in the crowd raised their hands to a drop kick by Morley that looked as if it had gone over – but the referee didn't think so. Then Aarvold broke clear but was stopped on the Wellington line by a last ditch tackle from Hazelwood. Aarvold got his pass away, but the Wellington defence managed to get the ball and scramble it away. The British closed the gap to 12-8 with a penalty goal, but desperate defence held them out to seal a famous victory.

At the final whistle the crowd rushed onto the field to lift their gallant heroes onto their shoulders. As an exceptional mark of appreciation the Wellington Union allowed the players to keep their playing gear.

27

THE BATTLE OF THE SOUTH

1 August 1936. Carisbrook, Dunedin

When 100 pipers were not enough

In 1935 the Shield dropped southwards like a stone. It started off in Auckland, but on 10 August Canterbury came to Eden Park and won 16-13. After four successful defences, Otago came to Lancaster Park for their 10th attempt to win it.

A crowd of 17,000 watched in perfect conditions as Canterbury took the lead with a penalty goal to Deavoll. Then 35 yards out from the Canterbury goal Trevathan was surrounded by opposing forwards but he managed to wriggle free and drop a goal that put the challengers into the lead 4-3. Otago attacked again through their forwards, then moved the ball out to the backs, enabling Watts to score in the corner. Right on the stroke of half time McAuliffe kicked a penalty goal for Canterbury bringing the difference back to just one point.

Canterbury had to come back in the second half if they were to retain the Shield but Otago were the speedier team both in the backs and the forwards. A breakaway by the Otago pack led to a try by Green. Even though the conversion took them out to 12-6 Otago remained on the offensive and scored again through Quaid to make the final score 15-6. Otago had won the Shield for the first time.

The following year the first challenge was from Otago's blood brothers – Southland. Six trainloads of supporters led by ten pipe bands totalling 100 pipers travelled up to Carisbrook. Heavy snow had fallen overnight and had to be scraped off the ground before the record 27,000 spectators could enjoy the spectacle.

Though Southland did much of the attacking, it was Otago who kept scoring the points with kicks and then a try to Watt in the corner bringing the score to 10-0. Southland's All Black Brushy Mitchell moved to centre in the second half and the result was that Southland scored with a try to Purdue. But a penalty goal and a second try to Watt finished off the southernmost rugby province. Southland were defeated but promised they would be back.

And they were. For the next 15 years the Ranfurly Shield did not move north of the Waitaki River as Otago and Southland jealously fought over it between themselves.

Six train loads of supporters led by ten pipe bands totalling 100 pipers travelled up to Carisbrook.

28

ALL BLACKS BEATEN BY SOUTH AFRICAN BASTARD

4 September 1937. Lancaster Park, Christchurch

Defeat in 1937 grated almost as much as defeat in 1905

It was all on again – the Rugbyweight Championship of the world. A contest over four rounds. In the right-hand corner wearing the black, the pride of New Zealand. In the left-hand corner wearing the green and the gold the touring South Africans. They were led by Philip Nel, who at 34 was ageing but still a powerful forward and an inspirational captain. The South Africans also had several other hard forwards including the flanker Ebbo Bastard.

In previous encounters honours had been even. Now one side was looking for a knockout. On the way over the tourists played 11 games in Australia including two torrid tests, neither of which was won with ease. They crossed the Tasman Sea aboard the *Wanganella* with some trepidation about the quality and the passion of the rugby they would encounter in New Zealand.

Their fears were somewhat unfounded, as they managed to beat all their provincial opponents, some by wide margins. However they knew that it was in the tests that they would be judged. In the first at Athletic Park, Wellington the Springboks were beaten 13-7. The second match in the three test series was crucial. If New Zealand won, that was the series.

On the morning of the game both teams attended a trotting meeting at New Brighton, leaving after the third race. Only then was the South African team announced. Key players were included who had been missing from the first test. They were Nel and fullback Gerry Brand, who had been sorely missed for his kicking. Danie Craven also moved to halfback from first five-eighth where he had not been happy in the Wellington match.

The second test was a hard, uncompromising match with the Springboks getting the better of the early forward exchanges and keeping the All Blacks in their half of the field. But then the All Blacks broke away after a dropped ball and were able to put Sullivan over from an overlap. Immediately hats, programmes and even coats were thrown into the air in joy.

Immediately hats, programmes and even coats were thrown into the air in joy.

It seemed the triumph was short lived as the Springboks came back up the field and threatened to score, but at the last minute Sullivan intercepted a long pass and raced up the field, punting over the head of Brand. The crowd screamed with excitement at the chase which Sullivan eventually won. At half time the All Blacks led 6-0.

In the second half the Springboks struck back with a converted try to Turner who brilliantly wrong-footed the defence by cutting back into the centre of the field. The forward exchanges became more vigorous and in the South African side both "Boy" Louw and Bastard played on despite being concussed. Desperately New Zealand sought to hold onto their one-point lead, but tiring, they gave away a penalty 40 yards out. The reliable Brand stepped forward and kicked it.

As the All Blacks fought to catch up again the Springboks continued the offensive. A pass was thrown out to the South African three-quarter Babrow and again it was intercepted – but this time by the Springbok Bastard who crashed over the line to score the try that put the game beyond doubt.

The South Africans went on to clean up the third test 17-6 in Auckland and win the series. It would be another 59 years before New Zealand would return the favour on South African soil. That was something that Ebbo Bastard would never live to see. He was later involved in a dispute with a neighbouring farmer. Both men were armed and shot each other to death.

29

THE KIWIS

24 November 1945. Twickenham, England

Where there are New Zealanders, there will be rugby...

When the Second World war was declared in 1939, sole All Black selector Ted McKenzie was told by the NZRFU to select a team for the 1940 tour to South Africa should the war be over by then. McKenzie selected the team, but sadly the war wasn't over and the names of those he had chosen went with him to his grave.

For the next six years official rugby came to a grinding halt. Yet rugby continued to be played by soldiers wherever there was a scrap of flat land, in the desert, in the jungle and even inside the barbed wire of prisoner-of-war camps.

Official service games were organised by General Bernard Freyberg (later to become Governor General of New Zealand) the Divisional Commander of the Second NZ Expeditionary Force in Egypt. Immediately after the war a New Zealand Army team, known as the Kiwis, was selected from troops in Egypt and Europe. Many players were specially flown to Britain from prisoner-of-war camps in Germany for the trials.

The captain selected to lead the Kiwis was Major Charlie Saxton who was rumoured to be the man McKenzie had chosen to captain the 1940 All Blacks. The Kiwis played an open style of rugby before huge and appreciative crowds who had been starved of top class sport during the war years. After being unbeaten in six matches the Kiwis were chosen to play against England at Twickenham in the first of the Victory internationals. The famous New Zealand commentator Winston McCarthy commentated half of the match for the BBC and the broadcast was relayed around the world to New Zealand.

In addition to Saxton at halfback, the Kiwis had Fred Allen (later to be All Black coach) at first five-eighth and the great Johnny (J.B.) Smith in the three-quarters. McCarthy said of Smith, "He had everything – poise, he could swerve, he could fend, he could side step off either foot, he could kick with either foot, he could change pace in a flash and above all, he never lost his cool."

The Kiwis were too much for the English and swept them away three tries to one, but it was a penalty goal by Herb Cook, who had displaced a young Bob Scott, that was perhaps the most famous moment of the match.

Cook was a strapping thirteen-and-a-half stone who stood just 5ft 7 inches tall in size four-and-a-half boots. Though he had seen Cook give the ball a mighty thump years before in New Zealand, McCarthy did not give him much hope of making this penalty which was placed on the halfway line just a yard in from touch. McCarthy told the listeners back in New Zealand as much, then added. "There will be a terrific roar from this vast crowd if it goes over. I"ll let you judge for yourselves how he goes. Here he is moving into the ball. He has kicked it and, LISTEN..."

Back in New Zealand those who had got up in the middle of the night to tune their ears into their radios heard a mighty roar as the ball sailed between the posts. The Kiwis won the match 18-3.

...rugby continued to be played by soldiers wherever there was a scrap of flat land, in the desert, in the jungle and even inside the barbed wire of prisoner-of-war camps.

30

MAORIS WALLOP THE WALLABIES

25 September 1946. Rugby Park, Hamilton

The first great victory of the Maoris

Maori rugby is unique and there are those who would say that the success of the All Blacks is based on the mix of the vibrant style of the Maoris tempered by European pragmatism to take the occasional kick for touch or goal.

A Maori XV played its first official match in 1910 when the team lost 11-0 to New South Wales in Sydney. Though in 1904 a Maori team defeated the Great Britain side in an unofficial match at Rotorua in the week leading up to the third test in Auckland.

Since 1910 the Maoris had played and lost official matches against South Africa, New Zealand, the British Lions and Australia. Though they had beaten France 12-3 at Paris on a tour in 1926-7, France were then considered the "easy beats of European rugby." In fact it has been said that in that game the Maoris taught the French how to play rugby. Afterwards the French adopted the Maoris "joi de vivre" slick passing style instead of trying to copy the more methodical "British" approach.

The captain of the team was "Doc" Paewai, a medical practitioner who stood just 5ft 6 inches tall but was a nuggety character who would tackle the biggest forward.

In 1946 the Maoris faced the Australian tourists at Hamilton. The week before the Australians had narrowly beaten a strong Wellington team by one point. The Maoris were weakened by the loss of their best player Johnny (J.B.) Smith through injury, though his brother Peter was in the three-quarters. Other notable Maori players were Ike Proctor who was said to have "hips like a working bullock" and Ron Bryers (an All Black in 1949) who was one of the best forwards in the country after the war. The captain of the team was "Doc" Paewai, a medical practitioner who stood just 5ft 6 inches tall but was a nuggety character who would tackle the biggest forward.

The match started quietly but the pace picked up after nine minutes when the Doc threw a huge dummy and scooted through for a try. A penalty goal took the lead to 6-0 at halftime. The Maori team increased the lead six minutes into the second half when Gardiner broke through the Australian defence and passed to Delamere who set up a try for Marriner. Isaacs converted.

Gardiner, who had only left school the year before, was having a superb game and created the next try for himself with a kick ahead. He followed up, the ball bounced kindly for him and he was able to dot down in the corner.

The experienced old Doc was also playing brilliantly and created an opportunity for Procter to put Smith away. He raced down the field to score. Smith and Procter were combining well together, more than making up for the absence of the more illustrious J.B. For the final try of the game Proctor again created an opening for Smith, who beat two men before scoring.

By this time the Maori team was running riot and the score would have been higher except for an excellent display by the Australian fullback Piper. When the whistle blew the score was 20-0 to the Maoris – a famous victory. Not only had they scored five tries against the Wallabies but they had kept them scoreless as well.

31

THE WARRIOR MAIDENS

28 September 1946. Carisbrook Dunedin

The first women's knockout tournament takes place

Ever since rugby was introduced into New Zealand in 1870 women had been itching to have a go. Why should the men have all the fun while the women were supposed to look after the kids and make the sandwiches for the aftermatch?

For years many women had been tempted to sneak into a male team in disguise, but were held back by the fact that not only would they have to look like a man and play like one, they would have to learn to swear like one too!

Just after the Second World War the women were given their chance in the male-oriented deepest south when a fully-fledged knockout tournament was organised for them. Four teams were entered – St Kilda, Dunedin Central, the Woodhaugh Wanderers and the daringly named Lovers Lane Ramblers.

"I've been waiting to assault someone for years, and today is my big chance."

The girls immediately made preparations. They informed their beaux that they would be "unavailable" on that particular afternoon for strolls around the park as they would be smack in the middle of it playing rugby.

The maidens soon discovered that there was no shortage of offers to coach them in not so subtle skills of this not so subtle game. In fact it was the first time in the history of rugby that a local coaching position had been oversubscribed. The St Kilda forwards were coached by Tommy Wallace, the Otago front row forward.

He no doubt instructed them in the hidden arts of front row warfare. There are rumours that his advice included, "Wear no perfume or make-up and don't shave on the morning of the match."

The girls were warned that there was to be no scratching or hair pulling. A light kiss on the cheek was acceptable after the game or to celebrate the scoring of a try.

Personal details were taken from the girls for insertion into the match programme, but none of them could be persuaded to divulge their weight and even if they had, it is doubtful if any of them would have been accurate. Before the match the referee also checked the girls for illegal length studs and hairpins.

Unfortunately the results of the matches have not come down the mists of time, but it is certain that a good time was had by all, that many mud packs were applied that day of the natural Dunedin variety and that the menfolk in the stand cut a good sandwich and poured a good cup of tea.

Nor is there any doubt that these novice women players had very swiftly acquired a comprehensive grasp of what the science of rugby is all about. As one player said before she took the field, "I've been waiting to assault someone for years, and today is my big chance."

32

THE MOST REMARKABLE KICK YOU EVER SAW

28 June 1947. Sydney Cricket Ground

Nutcracker takes his chance

There is a saying in rugby (or should be) that forwards are just frustrated backs. They know they can run with the ball better and could probably kick it better too if given the chance. One of these is Zinzan Brooke who takes drop kicks from anywhere and claims that when he was kicking the goals for his club side in Italy he had a better success rate than the great Naas Botha, who was also playing in Italy at the time.

But Brooke was preceded by another great back row forward who could also kick goals. Neville "Nutcracker" Thornton was a number 8 who was as fast as many of the backs he played with and he could kick with either foot. In one match for King Country against Hamilton before the war, he scored 17 points including a try, a conversion and 4 penalty goals. When he played for the 1945/6 Kiwi army team he drop-kicked a 45-yard goal against the French XV at Bordeaux.

After the war he represented Auckland where he took up a position as a schoolteacher. In 1947 he toured with the All Blacks to Australia. In the second test New Zealand were leading 24-14 with not long to play when the referee penalised Australia on the half-way line. With the match almost certainly won, the All Black captain Fred Allen called up fullback Bob Scott to kick the ball into the safety of touch.

But Scott wasn't too keen. He had a poisoned groin as a result of striking against the rock-hard Australian grounds. He had already converted all of the three tries that the All Blacks had scored as well as three penalty goals. He felt he had done his duty for the day and asked to be relieved of the chore. Immediately Thornton appeared and offered to take the kick for him.

Allen nodded, then turned to help attend the Australian fullback Dr Clem Windsor who had been hurt in the previous passage of play. Suddenly Allen turned back again in horror to see Thornton had placed the ball on the half-way line ready to take a kick at goal. According to radio commentator Winston McCarthy this is the dialogue that followed.

"Nut, what are you doing?"

"I'm going to kick for goal," came back the reply.

"I wanted you to put it into touch."

"Too late now. I can't touch it now – I've placed it."

Thornton was right. Allen wasn't too happy about it, but he had no choice. He wandered away grumbling.

Bob Scott sidled over and whispered to Thorndon in a low voice. "I bet you hit the corner post Nut."

"Watch me," said Thornton and backed right up to the 10-yard line. There were hoots and cries from the crowd as he began his run. According to McCarthy, Thornton seemed to ride a bike half way through his run and then he let fly with his boot. The ball shot off at a low trajectory and the laughter in the crowd died as the ball started to rise. Then as the ball rose higher they began to roar. The roar grew louder as the ball crossed two feet over the crossbar and it was still rising!

It went straight over the dead ball line and hit the back fence on the full. By that time the crowd had gone berserk, the great Bob Scott was standing watching in awe and the now smiling Allen was slapping the tall forward on the back.

The ball shot off at a low trajectory and the laughter in the crowd died as the ball started to rise. Then as the ball rose higher they began to roar.

33

CAVANAGH'S TALK

20 August 1947. Carisbrook, Dunedin

Those remarkable words by a remarkable man are etched into rugby folklore

For almost half a century the Otago team was coached by a Vic Cavanagh. There were two of them, father and son, both with the same name. They were referred to together as "the twa Vics" or individually as "Old Vic" and "Young Vic." When Otago won the Shield in 1936 Old Vic was the coach and Young Vic his assistant. Old Vic had retired by the time Otago won the Shield again, from Southland in 1947, and Young Vic was the coach.

Young Vic was undoubtedly one of the finest coaches *never* to coach the All Blacks. He had a profound influence on New Zealand rugby including the invention of the ruck, a formation of players so close together that you could throw a blanket over them, staying on their feet and driving over the ball to provide fast service to their backs.

As soon as Otago had won the Shield in 1947 Northern Invaders came hunting for it. First North Auckland (now Northland) were repulsed 31-12, but then came mighty Auckland with six All Blacks in the side including the captain Fred Allen.

...choosing his words ...then unleashing them in cutting phrases that turned the ordinary forwards of the first half into demons in the second.

An enormous midweek crowd of 25,000 turned up in brilliant sunshine to watch the spectacle. The packs immediately tore into each other with the Auckland forwards gaining early supremacy. It seemed the challengers were going to steamroller the holders out of the way. Deas scored two tries, Scott converted one of them and Ewert added a goal. The response from Otago was a paltry penalty by the fullback Haig. At half time Otago trailed 3-12.

No one really knows for sure what happened in the interval or the exact words that were spoken by Cavanagh to his "beaten" team. Cavanagh could not remember them afterwards. The players themselves were more aware of the intent than the content.

All sorts of rumours circulated later. That a shotgun was introduced into the conversation. That Cavanagh screamed at the players hysterically or alternatively, that he just stood and stared at them in glowering silence.

The truth was that Cavanagh pinpointed why Otago was losing (overconfidence) indicated the Achilles heel in the Auckland performance (their forwards were packing too high and too loosely) and outlined the method to beat them (full-blooded rushes into the Auckland forwards to demoralise them).

It was not just what he said but the way he said it, with smouldering presence and penetrating stare, choosing his words with deliberate care, suspending them momentarily in loaded pauses, then unleashing them in cutting phrases that turned the ordinary forwards of the first half into demons in the second. The players went out with coal in their hearts and fire in their blood. They subdued like a tame dog the Auckland team that had been so dominant in the first half. Auckland did not manage to score another point as Otago blasted them off the paddock to win 18-12.

It was one of the most remarkable turnarounds in rugby history and all because of a few words from a remarkable man.

34

BULLER BULLY THE BIG BOYS

30 July 1949. Carisbrook, Dunedin

In the days when tough guys worked with their hands

At the beginning of the 1949 season Otago had held the Shield for two years, but as is both the reward and curse of Shield-winning teams, eleven of their players had been selected to go with the All Blacks to South Africa. However the selectors made a major mistake in not selecting the Otago coach as the All Black coach. Perhaps there were some politics involved.

Instead Cavanagh had to stay behind and rebuild his decimated Otago team into one that could retain the Shield for another season. The first challenge seemed to be the easiest. It was from their country cousins from the rugged West of the South Island – Buller.

Today Buller is the smallest union in New Zealand rugby with just a handful of active teams and they usually sit in the lower end of the table of the third division of the NPC (National Provincial Championship). But in 1949 the Buller team was still augmented by miners and loggers, men who were hardened and toughened by manual work and who were not afraid to front up to the big boys.

As soon as [the Otago] backs had received the ball they found themselves being hunted down by the mad dog pack of Buller forwards.

The heavy ground conditions (though the weather was fine) suited them and before a crowd of 25,000 the Buller forwards immediately began to take the match to Otago. Led by Bill Mumm, who had an outstanding game, they packed well in the tight and hounded Otago in the loose. Even though Otago had the advantage in the lineout, as soon as their backs had received the ball they found themselves being hunted down by the mad dog pack of Buller forwards.

The Otago backs were too classy for the Buller backs, but even here there was no easy ride for the holders as they defended grimly every time Buller attacked. However it was Otago who scored first, six minutes into the game. Fitzgerald broke through the Buller defence and passed to McDonald. When McDonald lost control of the ball it was fly hacked on over the Buller line and McKenzie won the chase to dot down. With a penalty goal by Bell 10 minutes later it looked as if the smaller union was going to be swept off the field.

But the Buller forwards renewed their efforts and after a valiant drive towards the Otago line, Cunniffe forced his way over to score. A short kick took Otago by surprise and Buller rapidly moved the ball out to the winger Hawes who scored. The conversion was missed and at half-time the scores stood deadlocked at 6-6.

In the second half Buller again went on the attack and as the last minutes of the game ticked away they fought desperately to score again and achieve the miracle. But it was not to be.

They returned home with only memories, but ones that would flare bright again years later when they told their grandchildren how they had gone to Carisbrook Cathedral and almost snatched the Shield, forcing mighty Otago to scramble to a draw.

35

ELVIDGE'S BRAVERY

1 July 1950. Athletic Park, Wellington

One of the greatest skills on a rugby field is courage

In 1950 the British Lions came to tour. As usual they were packed with Celtic Wizards from England and Ireland. One of them, the Irishman Jackie Kyle, described as a "genius," was highly skilled in creating openings for the backs outside him. Another Welshman Ken Jones was a wing of incandescent speed who had reached the semi-finals of the 100 metres at the 1948 Olympics.

That this was a strong Lions side was shown when they drew the first test match with the All Blacks 9-9 at Dunedin. In that game both Kyle and Jones scored for the tourists. Only a try by Ron Elvidge, the New Zealand captain, pulled the home side out of the Lion's jaws of defeat.

The All Blacks had lost all six internationals they had played in 1949. As they came into the second test that meant it was now seven matches and two years since they had last tasted victory. On a Lancaster Park that was heavy from rain the All Blacks finally managed to grind their way to an 8-0 win in the second test. If they won the third test they would take the series.

The match was eagerly anticipated and people came from all directions. Most notable was a three-hour trip in wind and rain over the Rimutakas by a Mr A Bousted in his motor-driven wheelchair. As it transpired, the match was more than worth the journey.

After keeping the All Blacks in their own 25 for the first five minutes of the game, the Lions scored first with a penalty goal. Bob Scott then had several attempts at goal that missed. In the 20th minute disaster struck for the All Blacks when prop Johnny Simpson's knee was injured in a tackle. Simpson was known as the "Iron Man" and he was one of New Zealand's finest props, but his knee was destroyed and he never played again. In those days replacements were not permitted in internationals. New Zealand was down to 14 men.

Seven minutes before halftime disaster struck again when Elvidge was heavily tackled by the Lions three-quarter Matthews. The two clashed heads opening up a gash that required four stitches. Elvidge had also injured his collarbone and was forced to leave the field.

The All Blacks were now reduced to 13 men. They were forced to move a man out of the pack and rearrange the backline. For the remaining minutes of the half six All Black forwards valiantly fought to hold off eight Lions forwards.

As the second half began it seemed there was no way 13 All Blacks could defeat 15 Lions. Then a minute after the restart a figure appeared on the edge of the touchline. It was Elvidge. A hush fell over the crowd. They could see that he was badly injured with one arm hanging loosely, but he came back onto the field.

His body was shattered, yet he meant much more in spirit to the beleaguered All Blacks. He took up position as a rover behind the backline, a sort of second fullback, probably trying to scare the opposition into believing he was of more use than he was.

Late in the game when the All Blacks surged up the field on attack Elvidge followed up. Henderson passed to Johnstone who, suddenly finding his captain outside him, passed the ball on. Ignoring the pain, Elvidge clutched the ball to his

A hush fell over the crowd. They could see that he was badly injured with one arm hanging loosely, but he came back onto the field.

chest and ran forward as the Lions fullback Billy Cleaver came up to tackle him.

Just beyond Cleaver was the line. Elvidge knew he couldn't outpace Cleaver, he knew he didn't have the strength to fend him off, so instead he gritted his teeth and dived straight through the tackle to score.

That was the only try and the one that secured New Zealand a narrow 6-3 victory in a match that will always be remembered for Elvidge's moment of bravery.

36

POSSESSION, POSITION AND PACE

16 September 1950. Eden Park, Auckland

J.B. reminds everyone how to play rugby.

Rugby tends to follow trends. Some of them clever and some of them not so clever. After the 1937 Springboks won a series in New Zealand with heavier forwards, it was the "fatties" who were selected for the 1949 tour to South Africa. By then the Boks had discovered a new menace in Hennie Muller who flew from the back of the lineouts to create havoc among the New Zealand backs. When the All Blacks returned to New Zealand the whole country suffered from the disease of "Hennieitis" causing backlines to lose their imagination and take the safe option.

One team that did not suffer from this was North Auckland and in particular their great centre Johnny "J.B." Smith. J.B. had not toured South Africa because he was a Maori. But he did tour with the 1946 Kiwis whose philosophy was "Possession, Position and Pace."

J.B. had plenty of the last as well as many other exquisite abilities. Bob Scott had no doubt he was the greatest midfield back of his era. "I say without reservation that I was in mortal fear of Smith every time I played against him...I had not the slightest idea of what he would do next."

A tackler confronted him. One sidestep and J.B. was gone...

At the time North Auckland were a collection of similar individuals. There was All Black winger Brownie Cherrington, affectionately known as "the Big Horse," wily Doc Paewai and also in the team was an 18-year-old called Peter Jones who would later become more famous than any of them.

It was a team of characters and eccentrics that came to play the might of Auckland at Eden Park in 1950. However it was Auckland who almost started off with an eccentric try when Crowley caught the ball at a lineout and dived over to touch down. Except it was a little too eccentric for the referee's liking because Crowley forgot to hang onto the ball.

The Northerners scored an actual try from a break by Beazley with Dunn dotting down. Then the big horse galloped over for the second. Auckland fought back with a try to Kelly and a couple of those boring penalty goals. North Auckland were not interested in them, they just wanted to run the ball. Auckland got the idea too and ran it back at them. Right over their line. Try to Crowley. Try to Gilmour. Now Auckland led 21-10.

It was about time J.B. started putting *mortal fear* into the opposition. He got the ball in midfield. As the tackle came in, he seemed to have slipped the ball away. The tackler let go and J.B took off, running 20 yards to the goal line. He also had a hand in the next play when Beazley scored. The kick brought the scores back to Auckland 21, North Auckland 20.

Cherrington thundered away to score another and give his team the lead. Another run by him brought a try to Dunn. Then he was off again bursting down the field in a thrilling 40-yard run for the left-hand corner of the field. Suddenly he switched with J.B. who was now heading for the opposite corner. A tackler confronted him. One sidestep and J.B. was gone, a perfectly timed pass to Dunn and North Auckland had won 31-21.

The referee's whistle blew and the crowd ran onto the field to pick up J.B. and chair him off. He knew how to play rugby. Don't worry about your weight or Hennie Muller or anything else, just get hold of that leather oval and run with it.

37

SHIELD GOES WALKABOUT

August – September 1950. Carisbrook, Dunedin – Fraser Park, Timaru

...the North Auckland union paid out 1,000 pounds to fly the team to South Canterbury.

Five matches, five Shield holders

The Ranfurly Shield is famous for the eras when one team has managed to hold onto it. From 1935 to 1950 two great teams, Otago and Southland, jealously kept it between them in the lower tip of the South Island. In August 1950 that domination came to an end when Canterbury visited Carisbrook.

On the Saturday prior to this midweek match Southland had beaten Canterbury heavily by 25-8, so nobody in Dunedin was too worried about this challenge. But the 12,000 spectators who had turned up to watch in the driving rain became concerned when Otago took the wind in the first half and failed to score. The red and black forwards were on the rampage that day. They rushed about the field like driven men, beating the Otago forwards at nearly every phase of the play.

In the second half, with the advantage of the wind, Canterbury was able to score two tries (one converted) without response from the holders. At 8-0 the Shield was on its way to Christchurch.

There the red and blacks looked forward to another long era. But it was not to be. On 2 September at Lancaster Park it was Canterbury's turn to be held scoreless by Wairarapa. Again the underdogs took the holders by surprise. Though the Wairarapa forwards were much lighter, they were more aggressive and mobile.

The challengers won the Shield by making the only score of the game. Fullback Mahupuku fielded a clearing kick on the half-way line and kicked the ball between the posts for the 3-0 victory.

Wairarapa's first defence was against South Canterbury on 16 September before a crowd of 6,000 at the Solway Showgrounds in Masterton. Again the Wairarapa pack was aggressive and mobile, but the South Canterbury backs were superior. The match was a seesaw affair. South Canterbury scored first with a try to Stoddart. Wairarapa responded with a drop goal to level the scores and then took the score out to 9-3 with a try and a penalty goal. The challengers fought back and at half time the score was 9-9.

In the second half the Wairarapa backs went onto the attack and scored through Parker. The kick was converted to give the holders a 14-9 lead. But South Canterbury surged back with a try to Grant. The conversion brought the scores level at 14-14. In a whirlwind finish the ball was lost at a lineout and Patrick of Wairarapa made the mistake of trying to keep it in play. Grant grabbed it and sprinted 20 yards to secure the shield for South Canterbury 17-14 with just two minutes remaining.

Perhaps sensing that after its South Island seclusion the Shield was beginning to enjoy an itinerant lifestyle, North Auckland issued a challenge and were given the last possible date. Convinced they would win and stung by an unlucky loss to the British Lions 8-6 earlier in the season the North Auckland union paid out 1,000 pounds to fly the team to South Canterbury. With a fine display both in the forwards and the backs North Auckland swamped their South Island opponents and with a final score of 20-9 ensured that the Shield, which had been firmly lodged at the bottom of New Zealand, would now spend the summer in the northernmost part.

38

GEORGE NEPIA FINALLY RETIRES

30 September 1950. Gisborne

Even after almost 30 years George still didn't hang up his boots

Who is the greatest rugby player of all time? That is a conversation that can go on long into the early hours of the morning and still not be resolved. In rugby, men are judged by their deeds and what others have said about them, but more than anything greatness is measured by longevity. There are Colin Meads' 133 games for New Zealand and Sean Fitzpatrick's record of 92 tests, but both of those are spring chickens compared to the career of George Nepia.

Fitzpatrick was 21 when he first played first grade rugby and played his last game aged 34 – a span of 13 years. Colin Meads made his debut for King Country at 18 and played his last first class game in 1973 aged 37 – a span of 19 years. George Nepia made his first class debut aged 16 years and 30 days when he played in a Maori trial match for East Coast Districts on 25 May 1921.

The following year, aged 17, Nepia was selected by coach Norman McKenzie to play for the Hawkes Bay team. After the match McKenzie told him, " Well, that is quite sufficient for you this year. I'll put you aside and perhaps next year you'll be more experienced."

McKenzie was as good as his word and Nepia played for the Ranfurly Shield-winning Hawkes Bay team from 1922 to 1926. During that time he played in all 30 matches for the 1924/5 Invincibles, fashioning for himself a reputation as a rock-like fullback. He was given the ultimate accolade by noted critic of that era Denzil Batchelor, who said, "When I hear others debating who will play at fullback for the Kingdom of Heaven versus The Rest I turn to stone. It is not for me a question of whether Nepia was the best fullback in history. It is a question of which of the others is fit to lace his Cotton Oxford boots."

"When I hear others debating who will play at fullback for the Kingdom of Heaven versus The Rest I turn to stone. It is not for me a question of whether Nepia was the best fullback in history. It is a question of which of the others is fit to lace his Cotton Oxford boots."

Nepia last played for the All Blacks in 1930 and went into a period of semi-retirement from rugby while he eked out a living on his farm in those amateur days. He returned to play again in 1935 touring through Australia with the New Zealand Maoris.

That year he also took the coin of league, playing for the Streatham and Mitcham club and also for Halifax. During that time he also played for the Kiwis league team against Australia. But Nepia was to return to rugby courtesy of a wartime amnesty and continued to play the game well into his 40s. On 30 September 1950 he lined up for his last first class match when captaining the Olympians against Poverty Bay at Gisborne. At 45 years and 158 days that made him the oldest player ever to play first class rugby in New Zealand.

The captain of the opposing team was also called George Nepia. It was his son. That was only the second time in New Zealand history when father and son have played in the same match. A photo of the two captains tossing the coin before the kick off shows young George looking very much like his father in his playing days, while George senior looks like little more than an older brother, very fit for a man of his age.

So while Fitzpatrick played first class rugby for 13 years and Meads 19, there was a span of 29 years between the time George Nepia first laced his *Cotton Oxford boots* in earnest and the time he finally hung them up as a player (though he still continued to wear them for several years afterwards as a referee).

39

TIME WAS UP ON THE CLOCK

18 September 1954. Lancaster Park, Christchurch

The most thrilling and enjoyable of times

Canterbury won the Shield again in 1953 by thumping Wellington 24-3 at Athletic Park. The following season their hold on the "log of wood" seemed to be loosening. They could only draw with Waikato and struggled against Taranaki. By the 9th defence of their tenure against Otago the crowd of 30,000 that turned up to watch the game wondered if the red and blacks were vulnerable.

Otago thought so and went straight onto the attack. Five minutes into the match Canterbury suffered a disaster when the team's fullback Kevin Stuart broke his left arm. D. J. Mayo had to come on for his first Shield match to replace him.

Canterbury suffered more problems as Otago dominated the match. Wilson scored two penalty goals, then H.L. Smith broke through the holder's defence to offload to Grant who scored in the right-hand corner. By half time the challengers had racked up a healthy 9-0 lead.

It was only in the second half that the holders began to play. The luck started to fall their way when Mayo's penalty kick at goal hit the post and then bounced over the crossbar.

With only a quarter of the match remaining Canterbury was still behind when McPherson tried a kick across to the left wing. The rolling ball was picked up by the winger Ross Smith who sprinted up the touchline and over to score the try that brought the scores back to just 6-9.

The conversion kick was never taken. But it didn't need to be...

Canterbury now threw everything they had at the challengers but were met with fierce resistance. The watching crowd began to grow anxious as the minutes ticked away and every attack that was launched was repelled with equal ferocity. Finally the hands of the ground clock reached the point where the time was up. Otago supporters yearned to hear the shrill pitch of the referee's whistle while Canterbury die-hards hoped for just one more chance.

All seemed lost when yet another Canterbury attack broke down and the ball came into the possession of Otago's three-quarter Grant. He tried to clear but his kick fell well short of safety, straight into the hands of Canterbury's captain Pat Vincent. He started a rush down the right-hand side of the paddock, passing the ball on to Hern who slipped it to McPherson.

Otago players desperately hurtled across the field to try and stop the attack. It was Mayo, the debutante, racing up from fullback who made the extra man. He took the ball at speed and the whole crowd rose and cheered as he raced past them down the touchline to score the try that tied the score.

They were so delighted that the Shield was saved that they rushed onto the field. The conversion kick was never taken. But it didn't need to be, the 9-9 draw achieved in such a dramatic fashion, was all that was needed.

Mayo the debut hero was chaired off by the ecstatic crowd who surely would have agreed when Pat Vincent described the match as, "The most thrilling and enjoyable of challenges."

40

THE SKINNER TEST

18 August 1956. Lancaster Park, Christchurch

A hard man in a hard game

Gloves on again. The 'Boks were back and this time it was serious. In 1949 an All Black Team had travelled to South Africa and been whitewashed 4-0. Before that in 1937 the Springboks had won the series 2-1 in New Zealand. Anyone would start to suspect that New Zealand were not the best rugby team in the world.

In 1956 the whole country was gripped by the fervour to "beat the 'Boks." New Zealand was lucky to win the first test in Dunedin 10-6, but they lost the second at Wellington 8-3. Worst still the All Black prop Frank McAtamney had been "broken" by the Springbok front row.

Some iron was needed to shore up the New Zealand pack. The selectors turned to Kevin Skinner, the most experienced front row forward in the country, even though he had "retired" the year before. Skinner was the former New Zealand amateur boxing champion and a man who stood his ground when the heat was on.

Later some South Africans were to claim that Skinner was chosen to soften up the Springbok front row. Skinner claimed that during the match he only responded in the time honoured way to illegal Springbok tactics. The exact truth is lost in the mists of time and in that murky no-go area that is the front row.

...coach Bill Morrison, "used psychology to build Kevin Skinner up into a near frenzy...I think Kevin half expected the Springbok props to come onto the field in cages."

Skinner was certainly chosen to stiffen All Black resolve. When he arrived at the training camp he said he found that,"some guys believed that if we didn't win the third test the whole world would collapse." According to All Black Bill Clark, in the week leading up to the third test, the coach Bill Morrison, "used psychology to build Kevin Skinner up into a near frenzy...I think Kevin half expected the Springbok props to come onto the field in cages."

The battle lines were drawn, even though Skinner had made friends with one of the Springbok props Chris Koch when he had toured South Africa in 1949. At the beginning of the tour Koch had even looked up Skinner's mother in Dunedin. But this was rugby and the unofficial championship of the world was at stake.

When Koch came through the first All Black lineout Skinner issued him a warning. When he came through a second time Skinner hit him. As the match progressed matters became more heated. At one stage after a punch had just missed Skinner near the touchline, a policeman who was standing on duty nearby asked him if he needed a hand. Another time Skinner went to ground in a ruck. A Springbok boot flashed near his head. Skinner got up again snarling, "You b......s can't even kick straight."

In the second half Skinner swapped sides of the scrum with Ian Clarke who was having a torrid time from the other South African prop Jaap Bekker. Bekker was a 16-stone rock who it was rumoured practised his scrummaging against goal posts and had once snapped one. Skinner was unimpressed when he heard that story. "If that happened, the bloody goalpost must have had woodworm."

As they set for the first scrum of the second half Bekker tried to "pop" Skinner. The scrum broke up in a flurry of fists. Bekker didn't try to pop him again.

The All Blacks won the match 17-10 with three tries and some booming kicks from the debutante Don Clarke but it was Skinner who provided the unyielding iron that gave him the opportunity.

For years afterwards in South Africa Skinner was branded the villain for his "flying fists." In New Zealand most held the view that Skinner was simply a man who refused to be intimidated.

41

I'M ABSOLUTELY BUGGERED

1 September 1956. Eden Park, Auckland

The All Blacks finally win a series against the 'Boks

The victory in the third test of the 1956 series set up the fourth and deciding test as the most crucial game in the All Blacks' history. If they won they had the series. If they lost they still would not have won a series, home or away, against the 'Boks.

In Auckland that week the tension was incendiary. "If anyone had struck a match near it, it would have gone up," said Kevin Skinner. Nobody was thinking about anything other than beating the 'Boks.

"We were all desperate to win," said Skinner's fellow prop Ian Clarke. "No quarter would be given; indeed we would win by any means at all."

On the eve of the match the *Auckland Star* carried a cartoon of a little boy kneeling by his bed praying, "Please God, let the All Blacks win tomorrow." That

NEW ZEALAND HERALD

Players are strung out behind Peter Jones as he heads untouched towards the posts to score his memorable try at Eden Park that effectively secured the first series victory over South Africa.

night 15,000 people queued outside the ground to make sure they could get in. They lit fires to keep themselves warm.

At 8 a.m. the gates were unlocked with the announcement over the loudspeakers, "Open the gates for the test match of the century." A huge crowd of 35,000 people were in the stands by 9 o' clock waiting impatiently for the match to start. As the long morning wore on, elsewhere Peter Jones readied himself, "I started building up a feeling of power and dedication in my body, prowling around the room like a caged tiger."

"As I walked back there were just about tears ...all right, to be frank there were tears in my eyes," he said later.

The All Blacks got on their bus at 12.30 and threaded the way through the crowds, not looking outside at the faces, just concentrating on the game. In the changing room Ross Brown looked nervously across at Kevin Skinner, his boxer's face swathed in Vaseline. "Don't worry Brownie, we'll look after you," Skinner reassured him.

"Black, Black, Black," the crowd screamed as the players ran onto the field and battle was engaged for the unofficial Rugby Championship of the World. At one lineout early in the game the ball was thrown to Peter Jones who misread the signals and failed to jump for it. At another lineout just inside the Springbok half five minutes into the second half Jones made up for that error.

When the ball went loose on the Springbok side Jones took off after it. He kicked it on and it bounced towards the South African Viviers. In Jones own words, "Viviers' eyes were bulging...I could have let him take it [the ball] and bowled him for a row of ashcans – or taken it myself."

Jones went for the ball, virtually snatching it out of Viviers' grasp. Then he headed towards open field and the South African line. "I stoked everything up. After about five yards I could see or sense two men coming after me from the left. I ran harder than I'd ever run in my life." Jones outstripped his pursuers and touched down under the posts. "I had a sudden fear that this would be like a nightmare and the referee mightn't have seen it, or he might have dropped dead away down the field."

But he hadn't and the roar was deafening when the referee awarded the try. As Jones got up he could see the crowd standing waving their arms and throwing newspapers into the air. "As I walked back there were just about tears...all right, to be frank, there were tears in my eyes," he said later.

New Zealand won the game 11-5 and the series 3-1. Afterwards an exhausted Jones was dragged out of the changing room to utter the immortal words into a live radio microphone, "I hope I never have to play as tough a game as today's. I'm absolutely buggered."

Only as he heard the laughter of the crowd below did he realise what he had said. The honest words of the big North Auckland fisherman articulated the relief of the whole nation.

42

LIONS GET THE BOOT!

18 July 1959. Carisbrook, Dunedin

When the oyster shells were exhausted, the Lions turned to the less formidable arsenal of trifle and jelly.

Winning within the framework of the rules

The 1959 Lions came to New Zealand well equipped to play running rugby. They had two outstanding wings in the ebullient 19-year-old Irishman Tony O'Reilly and the elusive Englishman Peter Jackson who had the full bags of tricks with which to beat opponents. However the Lions had a shock in the first test when they ran into a human howitzer called Don Clarke.

NEW ZEALAND HERALD

Don the boot sends another kick spiralling towards the posts.

For 20 minutes the game was even, then the Lions were penalised for a scrum infringement 40 yards out. Up stepped Clarke, bang went the right boot. 3-0.

Two minutes later a Lion was caught offside from 35 yards out. Clarke stepped forward once more. 6-0.

The Lions' first try came when Risman created an opening for O'Reilly to go over in the corner. Then just before half time Risman kicked behind Clarke and Price won the race to the ball. At the break, including a penalty kick, the Lions led 9-6.

The Lions' backs were soon into the action in the second half. Evans intercepted and the ball was moved out to Jackson who scored in the corner.

Clarke missed two long-range penalty goals but then landed a third closer to the posts. 12-9.

The Lions ran the ball again after a knock-on by McMullen. Risman kicked and the ball was gathered by Scotland who went around Walsh and passed it to Price who scored his second try and the Lions' fourth. Risman converted to give the Lions a 17-9 lead.

The All Blacks still hadn't scored any tries and looked well beaten.

Then Clarke landed a 50-yard penalty. 17-12.

Then a 45-yard penalty. 17-15.

And finally a 35-yard kick to give the All Blacks an 18-17 lead.

By now the crowd had had enough "Red, Red, Red," they shouted in disgust. Many New Zealanders were embarrassed to win a test match without scoring a try when the opposition had scored four. The President of the NZRFU Mr Gordon Brown tried to make light of the situation at the dinner afterwards when he said the match had been won "within the framework of the rules."

The Lions were not impressed by Brown's words or Clarke's howitzers and started firing some of their own across the dining room in the form of the Southland oysters they had been served with. They were almost as lethal as Don Clarke's kicks, one of them drawing blood from an official who required stitches. When the oyster shells were exhausted, the Lions turned to the less formidable arsenal of trifle and jelly.

The hero of the day (or villain, depending on the point of view) was not impressed by the Lions' behaviour or their complaints. Clarke said,"We had them under pressure for most of the second spell and so were positioned to kick the goals when the penalties came."

But if only the All Blacks had scored just *one* try.

43

WHAT A THRILLER

28 September 1960. Eden Park, Auckland

Quick thinking by Nathan, slow thinking by Tremain

The match was almost over when Tremain lined up for another lineout. That was all he remembered. They carried him off on a stretcher.

Yet another exciting clash between Auckland and Canterbury took place in September 1960. This was Auckland's fifth defence after having snatched the Shield back from North Auckland on 31 August (to whom they had lost it just 11 days earlier).

Though the match against Canterbury took place on a Wednesday it attracted a huge crowd of 30,000 in perfect conditions. Playing for Auckland was Wilson Whineray who had played for Canterbury's Shield team of 1956. In the Auckland team Kel Tremain had also previously played for Canterbury. At the time he was temporarily living in Auckland and had been there long enough to qualify for the last three Ranfurly Shield matches of the season.

Auckland began the scoring when Tremain charged down an attempted clearing kick by Watt near the Canterbury line. He followed up and touched it down. Cormack converted which was the beginning of an artillery contest between him and the Canterbury kicker Henderson. An exchange of penalties between them ended up in an 11-9 lead to Auckland two minutes into the second half.

With 12 minutes of the match remaining Canterbury launched a raid from which Henderson scored to take the red and blacks into the lead for the first time. He tried to extend it further with the conversion, but his kick hit the post. It was to be a vital miss.

Three minutes later Auckland had the lead again with a penalty by Cormack. Now they were ahead by two points. It didn't last long. Four minutes later Henderson slotted over another penalty goal. Now Canterbury led by just one point. Two minutes later another penalty by Henderson took the score out to 18-14. As the minutes ticked away there was even more drama to come.

During the match Tremain had an ongoing tussle with his old mate the uncompromising Canterbury lock Tiny Hill. At first Hill couldn't work out why he was having difficulty jumping at the lineouts. Then he felt a tug, turned around and realised Tremain was holding him down. Hill offered the statutory warnings (two in this case as they were old friends) to which Tremain replied, "Just try granddad."

The match was almost over when Tremain lined up for another lineout. That was all he remembered. They carried him off on a stretcher.

Now just one minute remained on the clock with a scrum down and put in to Canterbury. It looked as if the Auckland were going to lose the Shield for the second time in 11 days.

Then Auckland won the tighthead. Herewini aimed a speculative kick towards the left-hand corner of the Canterbury tryline. Desperately players from both teams hared after it, but it was an awkward bounce. Hands stretched towards the ball and missed it. The ball rolled on into the Canterbury in-goal area. Then the Black Panther, Waka Nathan, pounced on it, but instead of touching down, he picked the ball up and coolly ran the 15 yards infield to score under the posts. The crowd let out a mighty roar.

That was what woke Tremain in the dressing room. He had no idea what was happening on the field.

There was another roar as Cormack placed the conversion that Nathan's quick thinking had made a formality. Auckland had retained the Shield at the last gasp, 19-18.

NEW ZEALAND HERALD

Kel Tremain has levelled the score against France at Athletic Park, but the most dramatic moment is yet to come as Don Clarke attempts the kick in a howling wind.

44

DADDY OF ALL KICKS

5 August 1961. Athletic Park, Wellington

Was the most amazing kick ever great skill or simply luck?

The French came to tour for the first time in 1961. They are very unpredictable in how they play their rugby. When they are good, they are very good. When they are not so good, they can be very bad, but...*c'est la vie,* as the French say.

There are those who say the Wellington weather is just as unpredictable, but there are others who say it is very predictable because it is always raining and there

The spectators in the stand could hardly believe it as they literally held onto their hats. Clarke ran up and kicked the ball almost *parallel* to the 25-yard line.

is always a wind blowing. In the second test against France at Athletic Park the wind was really blowing – at 140 kilometres per hour.

At sea the liner Canberra couldn't make it through Wellington heads, ferry sailings were postponed, Wellington airport was closed, most northern roads were blocked... but the weather was still not bad enough for the NZRFU to postpone the game. Thirty-five thousand spectators put their heads down and pushed their way through the buffeting wind to arrive at the game. Another 20,000 who had been expected listened to the roar of the wind rattling against their windows and decided to stay at home.

At the game the shrieks of the spectators could just be heard above the wind. Not at the rugby, but at the wind snatching the felt hats off their heads. The French won the toss and decided to play with the direction of the gale. It was so bad that when Don Clarke took his first 25 drop out in the normal way, the wind took hold of the ball and embarrassingly tossed it back over his head.

In the first half the All Blacks tactics were simply to deny the French the ball. When they got it the French were likely to kick, so John Graham was detached out of the pack and recruited as a second fullback to help Don Clarke field these gale force punts. At the lineouts the All Blacks threw every ball to Nev MacEwan who was standing right at the front. New Zealand survived the first half 0-0.

There were just 15 minutes left in the match before the first try came, but it went to France when they intercepted a loose New Zealand pass to score. The All Blacks attacked from the restart and almost scored. Then the French had a scrum near their own line. The ball came back to fullback Lacaze and he went to clear, but Tremain came roaring through and charged him down. Tremain was first to the loose ball to score.

Clarke was given the job of taking the conversion with "the wind roaring like a cyclone at my back." He had already missed two shots at goal in these impossible conditions. He took the ball back to the 25 and lined up to kick. But what was he doing? The spectators in the stand could hardly believe it as they literally held onto their hats. Clarke ran up and kicked the ball almost *parallel* to the 25-yard line.

The ball shot off on a straight trajectory, heading for the opposite stand, then suddenly the wind caught it and tossed it like a rag doll between the posts. The French gave a Gallic shrug of disbelief. Everybody said it was a miracle, but the man who took the kick preferred to describe it as a fluke. Fluke or miracle, this banana-shaped kick won the test.

45

DRAMA IN TIMARU

15 August 1961. Fraser Park, Timaru

Grandmother slugs it out with Attila

The French were involved in another remarkable match two weeks after the gale test when they took on South Canterbury who had never beaten an overseas team. Twenty-two thousand local rugby fans turned up to watch a match that was at times in danger of getting out of control.

One of those fans was 56-year-old grandmother Hilda Madsen who came to Timaru to watch the game alone with her husband, two sons and a nephew. Little

TIMARU HERALD

This South Canterbury forward doesn't look too worried about his opposing Frenchman. Perhaps he knows the battling grandmother is on her way.

...the police later declined to charge Hilda Madsen on the basis that not only would it have been ridiculous to have done so, but it would have been seen to be ridiculous.

did she realise that day she would become one of the most famous characters in New Zealand rugby.

The French could not have expected too much of a problem at Fraser Park. South Canterbury were one of the minnows of New Zealand rugby and after losing narrowly to the All Blacks at Wellington the Tricolours had beaten the powerful provinces of Otago and Southland by clear margins.

But the South Cantabrians came at them with a well-controlled fury which only increased when the Southerners were reduced to 14 men (replacements were not then allowed at international level) after the French flanker and captain Michel Crauste crash-tackled the South Canterbury first five-eighth Richardson out of the game. Crauste was a fine and aggressive forward who went by the self-explanatory nicknames of "Attila" and "Mongol".

But the committed South Cantabrians took the score out to 11-0 causing the temperamental French to overheat and the match became even more torrid. The French had caught up to 17-14 with the minutes ticking away when the South Canterbury second five-eighth Eddy Smith made a run down the touchline that would surely seal the game.

Standing nearby, shouting excitedly with the rest of the crowd, was Hilda Madsen and her family. She watched with shock as Smith was downed by a French fist and then the French flanker Michel Crauste moved towards Smith, as if he intended to do further damage. He picked him up, then dropped him again as South Cantabrian Ray Vercoe approached menacingly.

But it was too much for Hilda. The grandmother strode meaningfully unto the field with her walking stick in one hand. But it was the other hand that she balled into a fist and used to strike the Frenchman in the back. Crauste looked around in shock as the crowd gave a roar of approval. Before the 56-year-old grandmother could do more damage to the French player, two policeman quickly moved onto the field.

After the match which South Canterbury won 17-14 for a famous victory, the police later declined to charge Hilda Madsen on the basis that not only would it have been ridiculous to have done so, but it would have been seen to be ridiculous.

Afterwards Hilda said, " I was so mad I did not know what I was doing. I hit him hard. But I don't think it hurt. I think he got a bit of a shock though."

When Hilda passed away in 1988, aged 77, she was famous as "the woman who hit the Frenchman." Afterwards the legend developed that she had used an umbrella. Though Hilda had been carrying a walking stick, she definitely hit Monsieur Crauste with her fist. It was a good clean blow.

46

WELSH CURSE BROKEN

21 December 1963. Cardiff Arms Park, Wales

Clarke knocks on wood twice to beat ancient Druids

No one in New Zealand could forget what had happened at Cardiff Arms Park in 1905. In 1924 the Invincibles had gone back to Wales to avenge that defeat. They had trounced Wales 19-0, but that game was played at St Helen's Swansea. The next two matches between the two countries were played at Cardiff Arms Park and the Welsh won both in 1935 (13-12) and 1953 (13-8). It seemed that a Welsh Druid had sprinkled magic water over the field to cast a hex on New Zealand.

But the 1963 All Blacks had little respect for the invisible qualities of magic. Many of them, like Colin Meads and Ken Gray, were practical farmers, hard driving forwards who were more concerned about graft and pressure than Welsh Wizards waving their wands. They believed that All Black iron would put a dent into softer Celtic bronze, though Meads did admit, "Cardiff Arms Park, its forbidding atmosphere, made the Welsh appear invincible."

The match started in the traditional fashion with the Welsh singing as if their whole history depended on it. There was some nervousness in the Welsh camp because Wales had finished last in the previous season's Five Nations championship. But the All Blacks had not proved invincible on this tour, having gone down to Newport 3-0.

New Zealand was awarded a penalty from the second lineout of the game and though it was 58 yards out, Don Clarke lined up for the kick. The Welsh watched with both dread and awe as the ball thumped against the post and fell away. Three minutes later he had another chance and goaled from 20 yards.

Then Clarke took another penalty kick, this time from 54 yards out. Again he struck the post. A shudder went around the stadium. Wood was sacred to the ancient Druids. Would the knocking of wood twice break the ancient curse?

Meads and his fellow forwards did not worry about that. They just made sure to keep the Welsh pinned in their own half, both to prevent them from scoring and to keep them within the range of "the Boot." Meanwhile the Welsh were muttering to each other, "For 'eaven's sake boyos, no more penalties."

"For 'eaven's sake boyos, no more penalties."

After the break the All Blacks drove the Welsh back 20 yards in one scrum and kept the pressure on throughout the game. Clarke had another shot at goal from

40 yards but he was suffering from a groin strain and missed it. It was up to first five-eighth Bruce Watt to take New Zealand out to 6-0 with a drop goal. The Welsh tried to rally but could only manage a drop goal attempt which went wide.

In the final minutes of the game with time running out, Meads charged into the Welsh halfback and captain Clive Rowlands. Meads walked away, but Rowlands stayed down before rising groggily to his feet. The crowd immediately began to boo and jeer Meads, the sound of it continuing until the final whistle.

It filled the air of Cardiff Arms Park with a strange jarring sound so unlike the normal rhythmic melodies of Welsh singing. But was it really directed at Meads? As the end drew closer and with it a 6-0 victory to New Zealand, was the discordant sound emitting from those 58,000 Welsh throats really an acknowledgement that the hex had finally been broken?

47

MEADS WALKS

4 Nov 1967. Murrayfield, Edinburgh

The second man to be sent off in an international is also a New Zealander.

Colin Meads played his rugby from the late 1950s through to the early 1970s and like the fossilised footprint of a dinosaur, he left his imprint behind him. On the rugby field he was a colossus, intimidating and unrelenting, the physical commitment matched by an inner fire that took no prisoners and allowed of no compromise.

Other players would later recall their bruising encounters against him which took place both inside and outside the framework of the laws. He was not blameless, nor did he fail to attract the attention of equally uncompromising opposing players, nor of referees (though there are some who claim that in his later years Meads was the finest referee ever to play the game).

During the 1967 tour to the British Isles the All Blacks made a quick trip to Paris. But it was no holiday as they defeated the French team at Colombes stadium 21-15. During that game Meads attracted the attention of a French forward and received a kick in the head.

As a result, in the next match against Scotland he wore a white bandage on his head which made his easily recognisable appearance even more prominent in the vigorous forward encounters. In the first half Meads was involved in an incident when the All Black fullback Fergie McCormick was trapped at the bottom of a ruck.

"Get to me," McCormick shouted as he was being kicked and Meads walked across several Scottish players to try and help him. But the Irish referee Kevin Kelleher saw that as dangerous play from Meads. He awarded a penalty and gave Meads a formal caution.

Towards the end of the match (which New Zealand won 14-3) Meads was in a ruck with one Scotsman hitting him and another trying to pull him out, when suddenly he saw the ball come clear along the ground. He darted forward and aimed a kick at it just as the Scottish fly-half Chisholm tried to grab it. Meads struck the ball which cannoned straight into Chisholm who went down.

The whistle blew and to Meads' astonishment he saw the referee pointing towards the dressing room. Meads took a couple of paces and stopped. He was stunned. He just couldn't believe it. As if in a haze he saw the All Black Captain Brian Lochore talking to the referee. Meads waited for the decision to be reversed. But even if he could have changed Kelleher's mind Lochore had been at the bottom of the ruck and had not even seen the incident. "You're off," the referee said tersely and Meads took the long lonely walk to the booing of the crowd.

"You're off," the referee said tersely and Meads took the long lonely walk to the booing of the crowd.

The dressing room was locked when he got there. Two Scottish bandsmen were hanging about the worst for drink. They looked at Meads' bandaged head and thought he had gone off injured. When the dressing room door was unlocked Meads sat on the bench alone and was overcome by an overwhelming sadness. He felt he had let everybody down.

For days, months and even years afterwards the debate raged on about whether or not Meads should have been sent of. But it was too late. He had been. Only the second man (after Cyril Brownlie) to be sent from the field of play in an international match.

48

BEEGEE IN AFRICA

25 July 1970. Loftus Versfeld, Pretoria

Giant sidesteps make inroads into apartheid

There is no relationship in any sport that has been as fraught with difficulties as that between New Zealand and South Africa at rugby. On the field neither will yield any ground. Off the field the history of their battles has been one of firstly ignoring and eventually being overwhelmed by the South African policy of apartheid.

That first reared its head after the Maoris match against the Springboks in 1921. Then the first New Zealand team to tour South Africa in 1928 omitted the great George Nepia because he was a Maori. There were some who said the policy not to tour South Africa with Maoris was out of respect for internal South African policy. Others said it was out of respect for the Maoris and the way they might be treated in South Africa. But it was strange that the New Zealand selectors should acquiesce. It meant they were never able to send their strongest teams to South Africa.

Attitudes began to change in the late 1960s until finally in 1970 the All Blacks were able to include Maoris on the tour. The first three Maoris to tour South Africa were Blair Furlong, Buff Milner and Sid Going. But the player who gave the South Africans the most trouble on the tour was not even a Maori. The All Blacks' teenage wing Bryan Williams was of Samoan heritage.

Williams had thighs like totara trees, blinding speed and an enormous sidestep. Years later when Williams tried to pass on that sidestep to John Kirwan, Kirwan found it impossible to repeat because Williams actually missed out an entire step when he did it.

Williams had thighs like totara trees, blinding speed and an enormous sidestep.

Williams was an early starter, making his senior debut for Ponsonby on the wing aged 16 and scoring a try. In 1969 he became a regular in the Auckland team. The following year he made the All Black trials and though he did not have any attacking opportunities was able to show the quality of his defence including twice running down opposing wing Bruce Hunter from behind. Both Williams and Hunter made the All Black team to South Africa.

In his first game for the All Blacks on that tour Williams scored two tries and still aged only 19 years and 295 days he was selected for the first test against the Springboks. South Africa came at the All Blacks with a fury Williams had never

NEW ZEALAND HERALD

Beegee in full flight, the ball tucked under his arm and the ground shuddering to the pounding of his legs like totara trees.

experienced before or since. The 'Boks fired out to a 12-0 lead at half time. In the second half the All Blacks fought back. Laidlaw was substituted by Sid Going who immediately made an impact with some devastating breaks. On one of these, midway through the second half, he came around the blind side of the scrum and passed to Williams who was 40 yards out. BeeGee surged forward, sidestepped the South African fullback McCallum and ran in his first test try.

Though the All Blacks lost the match 17-6, this was just one of the 14 tries Williams ran in during the tour. By the end of that tour Williams would be rated the greatest winger in the world and receive more than 1500 letters from his South African fans. Both on and off the field, the colour bar had been quite definitely broken.

49

IRON PINE TREE

19 August 1970. Boet Erasmus Stadium, Port Elizabeth

Colin Meads needed a lot of convincing that he had a broken arm

Has anybody ever played rugby who was tougher than Colin "Pinetree" Meads? He was raised on a rugged King Country farm and played at a time when the main painkiller used was the decision not to feel the pain. Such as in the 1965 test against South Africa at Eden Park where Colin Meads' brother Stan was felled. As he lay on the ground Colin's shadow fell over him.

"How are you?" Colin asked.

"No bloody good," Stan replied.

"Well, you'd better get yourself right," Colin said tersely. "In one minute's time you and I have got to lock a New Zealand scrum together and we're not going backward. Get yourself right." Colin strode away and Stan dragged himself back to his feet.

When he first became an All Black Colin Meads had served an unofficial apprenticeship under All Black lock Tiny Hill. Hill played matches dressed in skull cap and shoulder pads, but claimed that these were only to protect the opposition, not his own body. "He was a ruthless player," Meads said of him later. "He was a player who hurt when he brushed against you." Meads learned to play in the same uncompromising style, sparing neither his opponents nor himself.

"How are you?" Colin asked. "No bloody good," Stan replied. "Well, you'd better get yourself right," Colin said tersely.

During the South African tour Meads had been playing at his peak when the All Blacks took on Eastern Transvaal. In that game Meads was trapped in a ruck with his arm exposed when suddenly he felt a sharp pain. He says that the man who drove his boot into the exposed limb was the Eastern Transvaal captain Skip Henderson.

Meads' arm was numbed with pain as he rose to his feet. Pain meant little to Meads. He played on. But he found he could not use his arm to hold the prop in the scrums. Captain Brain Lochore told him to get it checked. Meads went to the sideline where the doctor examined him. He advised Meads to play on.

Only later did X rays reveal that he had played the remainder of the match with a broken arm – one of the two bones in the forearm had been fractured. The medical prognosis was that Meads could be out anything from two weeks until the

Brian Lochore has the ball but the eyes are immediately drawn to Meads, anticipating every move.

end of the tour. Nine games later he was back, the arm still not healed but reinforced by a leather case. Meads played well enough and the All Blacks won against South-Western districts at George.

But would the arm be strong enough for the cauldron of test match rugby? Having missed the first two tests, Meads was selected for the third. It was the sporting story of the year. Here was the iron man who would not lie down, the man so tough that he played a test match, his 50th, with a broken arm that was still not fully healed. Though the match was lost 14-3 no one would forget Meads' selfless courage in taking the field.

50

GOING, GOING GONE!

1971. Eden Park, Auckland

No brotherly love in Shield battle

During the latter half of the 1960s the Ranfurly Shield was largely the province of the Hawkes Bay Union. They won it in 1966 in the last Shield game of the year defeating Waikato 6-0 in Hamilton and held it until the very last Shield game of the decade when Canterbury snatched it off them 18-11.

In the second defence of 1971 (having resisted eight challenges in 1970) Canterbury finally succumbed to the northern challenge, losing 20-16 to Auckland. Auckland's first defence was against North Auckland.

The region that stretches like a finger and points New Zealand at the rest of the world has produced some interesting characters over the years such as J.B. Smith, Brownie Cherrington, Peter Jones and more recently Norm Berryman, but no name is more synonymous with the region than that of *Going.*

North Auckland (or Northland as they are now) never seems to go a game without at least one Going on the team sheet. It started back in the 60s with brothers Sid, Ken and Brian.

"You kick it, you gave away that try."

Sid was the most famous of the three, a darting fireball who represented North Auckland at halfback for 14 seasons and New Zealand in 29 tests. Nobody has ever played the game who was better at scorching the grass down the blind side of the scrum than "Super Sid."

Going was a mercurial fireball of a player. Here he opposes another great halfback, Gareth Edwards, playing for the Lions.

Ken, a fullback, also represented the All Blacks, playing three matches during the 1974 season. The third member of the trio, Brian, gave long service to North Auckland, making his debut in 1967 and playing his last game in 1978. The brothers had an extraordinary understanding between each other and worked many slick combinations both for North Auckland and NZ Maoris.

The understanding wasn't quite so good though in 1971 when North Auckland came to Eden Park to challenge for the Shield. In a fiercely fought contest it was a try to "Super Sid" that put the challengers ahead. Then North Auckland got a penalty. Sid tried to ram home the advantage by taking a quick tap kick, but the opportunity was muffed. As a result Auckland got hold of the ball and scored.

As the North Auckland team lined up behind the posts for the conversion, the elder Ken was furious with Sid for giving away the try. He was so incensed that

North Auckland's All Black prop Richie Guy (who later became Chairman of the New Zealand Rugby Football Union) had to intervene to prevent Ken physically attacking his brother.

The play moved upfield again and North Auckland was awarded a penalty on half way. It was a long kick, but Captain Sid reckoned kicker Ken could make it. Sid tossed him the ball.

Ken tossed it straight back.

"You kick it, you gave away that try."

Sid placed the ball, ran up and kicked it over. The brothers went back up North with the Shield that day. They say that if you are ever in the region and hear a distant rumble of thunder but no rain falls, it is only the Going brothers still arguing about that Shield match at Eden Park.

51

KIRKPATRICK'S TRY

10 July 1971. Lancaster Park, Christchurch

One of the greatest individual tries of all time

The 1971 Lions were one of the finest teams ever to visit New Zealand. They had virtually no weaknesses, containing an explosive mix of Celtic backs in Gerald Davies, Barry John and Gerald Davies (Wales) and Mike Gibson (Ireland) along with a forward pack of real commitment and scrummaging power.

Kirkpatrick suddenly broke from a maul with the ball in hand and proceeded to fend off player after player on a scintillating sprint to the try line.

In the first test at Carisbrook the Lions had won 9-3 and the teasing kicking game of Barry John had effectively ended the international career of All Black fullback Fergie McCormick. The All Blacks were the underdogs for the second test at Lancaster Park after the Lions had continued on their winning way with clear victories over Southland, Taranaki and New Zealand Universities. But it was a game the All Blacks had to win.

From the start they attacked through the pack with feisty Northland halfback Sid Going almost acting as a ninth forward. Five minutes into the match the pressure told when Going fed Burgess from a scrum and the first five-eighths dodged through the Lions defence to score. The Lions responded 16 minutes later when their fullback John Williams caught the ball on his own 22. He fed Gibson who burst upfield and drew Mains to send Gerald Davies on a sprint over the line.

Gerald Davies then "scored a try" for New Zealand when he was guilty of tackling Brian Williams before Beegee had received the ball and the referee awarded a penalty try. Going made another break, again feeding Burgess who again scored in a virtual repeat of the earlier try.

Though the All Blacks were ahead, the Lions had plenty of firepower in their team with which to mount a comeback. One more New Zealand try was needed to finish them off and what a try it was.

In the All Black side that day was Ian Kirkpatrick, one of the most explosive flankers ever to play the game. The Poverty Bay farmer stood 1.89m tall and weighed in at 103kg. In his younger days he had been outstanding at athletics, but gave all that up to concentrate on rugby.

As the Lions sought to mount an attack from the half-way line, Kirkpatrick suddenly broke from a maul with the ball in hand and proceeded to fend off player after player on a scintillating sprint to the tryline, moving in a diagonal path towards the corner flag.

The last line of defence was perhaps the greatest fullback in the world at the time, J.P.R. Williams. Williams was determined to make the tackle and Kirkpatrick was equally determined to score. Three times Kirkpatrick was forced to fend off the tenacious Welshman, stumbling as he made the final fend, but possessing the strength to drive himself over the line for one of the greatest tries ever seen on a rugby field.

52

GHOSTS OF THE PAST HAUNT WALES

2 December 1972. Cardiff Arms Park, Wales

Welsh stumble under the weight of their own heritage

"You never beat the Welsh, you just score more points than them," so the saying goes. The All Black team to play Wales in Cardiff on the 1972 tour knew they were not going to get an easy match. They might have won the last two encounters on the hallowed Welsh turf in 1963 and 1967 but many of the Welsh had been part of the Lions team that had won the series in New Zealand the year before.

If they needed any reminding about the passion of the Welsh, it happened on a chilly night at Stradey Park, when the Llanelli club side downed the mighty All Blacks 9-3, an event that was recorded for posterity by the modern Welsh bard Max Boyce in his song, *The Night the Pubs Ran Dry.*

Welsh rugby was entering another of its golden ages. It was true the team had lost "The King" Barry John who had retired, but the mythical Welsh fly-half factory already had an heir waiting – Phil Bennett.

Despite their great heritage though, the Welsh had not beaten the All Blacks since 1953. This was too much for the passionate Welsh to bear. But rather than wait for a mythical Arthur and his knights to appear from the mists of Avalon, instead they called on the real legends of the past.

Gnarled old warriors from previous decades made the pilgrimage to Cardiff Arms Park, some with the aid of walking sticks, many with bulging ears and flattened faces and bodies that had been bent and squashed in pursuit of rugby glory for the "Land of My Fathers". Their cracked old voices spoke of epic victories, long gone, but that would be revisited that day. That *had* to be revisited.

By comparison the All Blacks were not expected to win. In their side was a 21-year-old fullback Joe Karam, who had little experience of the spitting furnace that is test match rugby. Karam had started the tour as understudy to Trevor Morris who strained his calf on the New York leg of the trip leaving Karam as the only fullback.

This was his first test but it was the Welshman Phil Bennett who seemed to be suffering from nerves at the kick-off, putting the ball out on the full. More nerves from the over-motivated Welsh meant their flankers were caught offside just two minutes into the game.

It was Karam's turn for nerves as he lined up to take the kick. His knees were trembling so much that he took seven steps for his run up instead of the usual two. "I was so wonky I thought I'd fall over," he said afterwards. If that kick had missed, God knows what would have happened; I'm sure I would have gone to pieces."

His knees were trembling so much that he took seven steps for his run up instead of the usual two. "I was so wonky I thought I'd fall over."

It went over and so did the next and the next as Karam and Phil Bennett entered into a kicking duel. Karam kicked five against Bennett's four. New Zealand scored a try when Going hosted a high kick in the Welsh half and managed to snatch up the loose ball. He passed the ball to Murdoch who crashed over for the try (Murdoch's glory was short-lived. Within a matter of days he had been sent home for striking a Welsh security guard).

The Welsh responded with a try when Going was caught in possession by Gareth Edwards. They kicked ahead and the ball was picked up by Bennett who passed outside to the powerful wing Bevan who ran the ball in from 40 metres. Now the All Blacks led by just 19-16.

The Welsh attacked with passion and fury, causing the All Blacks to concede a penalty. Phil Bennett went to take a quick tap, but the Welsh Captain Delme

Thomas overruled him. Instead he instructed Bennett to kick for goal and level the scores.

The Welsh held their breath as the ball was kicked. It rose towards the posts and passed just outside the upright. A moment later the referee's whistle blew. In the crowd many old Welsh players clutched their fingers white to the knuckles around their walking sticks before bowing their heads and shuffling sadly away.

53

ALL BLACKS BEAT SCOTLAND AT WATER POLO

14 June 1975. Eden Park, Auckland

The only All Black Test of 1975 took place on the wettest day of the year

Before the First World War, Eden Park in Auckland was a lake crossed by a causeway. The land was reclaimed and a park created on which rugby posts were erected. The ground hosted its first rugby test between South Africa and New Zealand on 27 August 1921 (South Africa won 9-5).

In the middle of June 1975 the weather decided to reclaim Eden Park for its original use. That coincided with the first tour by Scotland to New Zealand. Scotland had several players from the 1971 Lions including the Scottish captain Ian McLauchlan and fellow forwards Sandy Carmichael and Gordon Brown. The Scottish tourists' first game was against Nelson Bays where they were made welcome by a pipe and drum band resplendent in kilts.

The weather was fine for that first match but had turned against them by the time they headed for Auckland for the final game of the tour and the only test. Heavy rain began to fall on the morning of the match and it just kept coming. By kick-off 80 mm had descended from the skies and it seemed that not a drop of it had drained off the surface of the ground. Eden Park had reverted to being a lake.

The weather was so bad that about 10,000 of the people who had paid in advance for their seats did not bother turning up. The spectators who did arrive were dressed in waterproofs or huddled under umbrellas or makeshift plastic tarpaulins which they sat under in groups and which the rain pelted down on like machine-gun bullets.

They peered through the rain as the All Blacks squelched out onto the field dressed in white jerseys which were drenched before they had even finished the haka. By this time a sheet of water covered the field and the conditions were quite farcical for international rugby. Almost every incident of play was accompanied by the visual effects of spraying water and there were rumours that players at the bottom of the rucks were in danger of drowning. There was also talk of dirty play such as deliberately splashing water into the faces of opponents.

The question of the match was not necessarily which team could play the best rugby, but which could best adapt to the atrocious conditions. The All Blacks had two advantages there in Ian Kirkpatrick who cleaned up the messy ball that came back from the lineouts and Sid Going who was good at cleaning up and running just about any ball.

Going had the added advantage of a receding hairline so that the rain just aquaplaned off the bare areas. New Zealand scored first when Going ran (waded) down the blind side from a scrum near the Scottish 22. Kirkpatrick burst away from the maul that formed and fed MacDonald who fell over the line, the waves of water that rippled to the corner flags acknowledging the try as much as the referee's whistle.

The All Blacks scored again in the second half when Steele failed to hang onto the slippery ball (who could?) after Bryan Williams had hoisted a high kick. It was Williams who dived for the loose ball to score. Williams scored a second after a breathtaking raking run (some supporters claimed afterwards that he had run directly *on top of* the water). Robertson had earlier made different use of the conditions, sliding through the water to score his try.

Everybody was happy when the referee blew his whistle for the end of the match. New Zealand had won 24-0, the spectators had seen some thrilling rugby despite the conditions and the Scots could look forward to warm showers.

And it was still raining.

54

CLASSIC CANTABRIANS

March 1976. Happy Valley Stadium, Hong Kong

NZ team wins the first Hong Kong Sevens

Sevens rugby originated in Scotland in 1873 when a canny Melrose butcher called Adam "Ned" Haig thought up a new way of creating a fund-raising tournament that would need fewer players and therefore attract more teams. The concept mainly remained confined to the border country of Scotland for a number of years until the Middlesex sevens were inaugurated in England in 1926.

Slowly the concept spread around the world. In the mid 70s it arrived in Hong Kong. When the British had acquired Hong Kong in the previous century it had been described as "A barren Island with hardly a house upon it." By 1975 Hong Kong had grown to become a thriving city of six million people and, like any English colony, it had a rugby club.

One spring day in 1975 the Chairman of the Hong Kong Union, a South African called Tokkie Smith, was scratching his head trying to work out the same problem that Haig had solved all those years before – how to raise some sponsorship money by holding a rugby tournament. What he came up with was a 15-a-side tournament featuring top teams from around the world.

A proposal was sent to HQ at Twickenham, but this was a full decade before even the first World Cup and the stuffy administrators there could see little virtue in the idea.

A proposal was sent to HQ at Twickenham, but this was a full decade before even the first World Cup and the stuffy administrators there could see little virtue in the idea.

But Smith still struggled on with the logistics of how to pull off his vision. He came up with the same solution as Haig had and decided to make it a sevens tournaments. Even with the sponsorship of Cathay Pacific Airways the cost to fly the best teams from all around the world was still prohibitive, so the tournament was restricted to teams from the Pacific Rim. By inviting clubs, rather than nations, the Hong Kong Rugby Football Union was able to sidestep the snooty ban by the gin and tonic rattlers of Twickenham.

The first game kicked off on 28 March 1976 between an Australian team, the Walleroos, and Korea. From the start the tournament was an immediate success with players and crowds alike. Both enjoyed the fact that the spectators were almost claustrophobically close to the action. Immediately the Hong Kong sevens developed a "village" atmosphere which has persisted to the current day.

Of the 12 original teams chosen, the New Zealand entry was the national sevens champions – the Cantabrians Club from Canterbury. They fought their way through to the final where they outplayed the Wallaroos 24-8, so that a New Zealand team became the winners of the inaugural Hong Kong Sevens.

55

LEAVE THE LAMBS AND PLAY RUGBY

12 May 1976. McLean Park, Napier

The first NPC game

Until the formation of the National Provincial Championship (NPC) in the mid-1970s fixtures between the different unions were arranged in a fairly casual manner at the beginning of each season. Local unions often had their own preferences. Taranaki tried to avoid playing in the calving season, while Southland liked the lambing season left free. Apart from that, the only real issue was who would challenge for the Ranfurly Shield.

An idea was first put forward for an interprovincial championship in 1972, but was not really considered until 1974. The initial proposal was for a top tier of 11 unions who would play against each other on a home and away basis, followed by two junior divisions, one consisting of the remainder of the North Island Unions and the other the bottom South Island unions.

Local unions often had their own preferences. Taranaki tried to avoid playing in the calving season, while Southland liked the lambing season left free.

At first the idea was slow to catch on. Rugby administrators tend to be traditionalists, but by the NZRFU AGM of 2 April support was beginning to gather for the scheme. A motion was passed in favour and a working party got down to the brass tacks of implementation assisted by sponsorship from Radio New Zealand.

Their first task was to make an enormous mathematical calculation of the results of all the provincial matches played over the last five years to see who would be in the top division. Those unions turned out to be Auckland, North Auckland, Bay of Plenty, Canterbury, Counties, Hawkes Bay, Manawatu, Marlborough, Otago, Southland and Wellington. It was decided that the bottom Division 1 North Island side would be automatically replaced by the top Division 2 North side, but that the Division 2 South champions would have to play off against the bottom South side from Division 1.

With all the paperwork out of the way, the first NPC match took place at Napier on Wednesday 12 May. Hawkes Bay took on Auckland before 10,000 enthusiastic spectators. The visitors had obviously made the long trip hoping they would record the first victory in the NPC and started full of fire and fury.

Auckland went ahead first with a try to their teenage winger Tua Saseve. Hawkes Bay came back with a try to Robbie Stuart. Then they equalled the scores with a penalty from fullback Roger Whinham who was making his debut. Hawkes Bay pushed ahead for the win, dominating the rest of the match. It would have been a famous victory for the country team beating the big boys from the city, but it was not to be.

Mark Sisam goaled from forty metres out to give Auckland a narrow and historic victory. They had been the very first team to be awarded the Ranfurly Shield and now became the first team to win an NPC match.

56

BUZZSAW BATTY CARVES UP LIONS

18 June 1977. Athletic Park, Wellington

An unconventional try from an unconventional character

The 1977 British Lions team was captained by Phil Bennett and included Irishman Mike Gibson who was the only surviving Lion from the 1966 Lions (Gibson had also toured with the 1971 Lions). Though this team had glimpses of the great Lions team of 1971, which won the series in New Zealand, and the 1974 team which remained unbeaten in South Africa, it was missing some key personnel from those sides – including Gareth Edwards and J.P.R. Williams. During the 1970s there was also a subtle but progressive change in emphasis in British rugby away from the exciting back play of the past towards a more forward-oriented approach.

The Lions had won all their matches coming into the first test, except for the match just before it where they had surprisingly gone down to New Zealand Universities.

They took an early lead in the test when New Zealand was ruled offside at a lineout and the Lions fullback Andy Irvine kicked the penalty. The All Blacks

NEW ZEALAND HERALD

Pocket battleship Grant Batty sets off on the run that ended in a '12-point' try against the 1977 Lions.

Like a flash Batty was onto it, intercepted and sprinted off down the field.

immediately attacked back. Sid Going got the ball from a ruck and broke. When he found his way momentarily blocked, he unexpectedly turned back inside to run virtually through the entire Lions pack to score.

The Lions took the lead again with two penalty kicks to Bennett, then Brad Johnstone followed up an unsuccessful penalty kick from the All Blacks to retrieve the ball and score. The conversion gave New Zealand the lead again.

They held it for just one minute. Going was penalised for not putting the ball into the scrum correctly and once more the trusty boot of Bennett pushed the Lions ahead. The match was so close that something had to give. Two minutes before half time it did.

The Lions halfback Brynmor Williams broke from a scrum inside the Lions half of the field. He passed on to Fenwick who pushed the ball out to the loosie Evans in support. Suddenly the All Blacks were outmanoeuvred and outnumbered. Four Lions players bore down on Grant Batty. A try seemed certain.

Batty was undoubtedly one of the most unusual and talented players ever to pull on an All Black shirt. He was small for an international player, standing just 1.65m high but he had a strong physique and a heart probably bigger than anyone who has ever played for New Zealand. He had a temper to go with it too and could best be described as feisty. The great Scottish commentator Bill McLaren said of him that he was like "a little buzzsaw."

But he was a good man to have around in a tough situation. Probably the Lions were aware of that as they converged on him in superior numbers and perhaps expecting a tough Batty tackle or something unpredictable they panicked and looped out an awkward high pass. Like a flash Batty was onto it, intercepted and sprinted off down the field.

When playing the year before in South Africa Batty had injured his knee and was forced to wear a specially designed metallic knee brace covered with padding. The knee still wasn't fully right and Batty would retire soon afterwards. What were the odds on him running the 60 metres to the Lions line without being caught?

Price and Irvine did catch up with him, but by that time Buzzsaw Batty was dashing the ball down beside the posts. It was the key moment in the All Blacks' 16-12 victory. Afterwards Lions coach John Dawes ruefully described it as a "12 point" try.

57

CHIPPIE HITS THE WOODWORK

12 September 1978. Palmerston North

A big scrum wins the day

The Ranfurly Shield is unique not only in New Zealand rugby, but also to any sport. It is a primeval contest without excuses or recourse where one side will walk away victors and the other will limp away as vanquished. The losers may never get another chance. And they know it, so they keep fighting until they hear the last blast of the referee's whistle.

The closest thing to a Ranfurly Shield match in any other sport is a world championship boxing bout where both men refuse to back down and continue to slug it out, digging deep to find the knockout punch as the last seconds of the last round tick away...

In September 1978 Manawatu held the Shield. They had a fearsome forward pack with several All Blacks and were expected to easily repulse the challenge of North Auckland. But the Northern men had issued a midweek challenge which is the most dangerous of all. The holders can become lulled into a false sense of security while the challengers are all fired up for what, for some of them, might be the biggest day of their lives.

Sure enough Manawatu were rocked back by the North Auckland assault. Thoughts of a comfortable home win were forgotten as Manawatu's rugged forwards realised they had some work to do. The teams battled it out, never one drawing away from the other, no one individual able to stamp his mark on the game to score the knockout try.

...those directly behind the posts knew exactly where it was going and could hardly bear to watch as the ball headed directly towards the left-hand post.

With nine minutes of the match remaining Manawatu were two points behind. The relatively small crowd of 11,000 breathed a sigh of relief as Jim Carroll kicked a remarkable drop goal to squeeze them just one point ahead. It would be enough – as long as North Auckland did not score again.

North Auckland went on the attack and Manawatu defended desperately. Now the home side was like an exhausted boxer holding in the clinches. They didn't care about scoring themselves, all they wanted to do was to eat up that most invisible and unrelenting of all factors – time.

Finally they did it. They had played 80 minutes of rugby. But the referee did not yet blow the game up. The Manawatu players asked how long there was to go. He told them there was some injury time. Whistles started in the crowd, eyes glanced up pleadingly towards the heavens.

There was relief as another North Auckland attack broke down. Manawatu had the scrum feed. Surely that would be followed by a kick straight into touch. As the Manawatu forwards set for the scrum, again they inquired of the ref. *Time up yet?* Again he shook his head.

The Manawatu players were slow to form, their halfback John Gilmer took his time putting in the ball.

As soon as he did, the North Auckland forwards gave a mighty shove, straining every muscle and sinew, hearts pumping and pushing tired bodies to the limit, they shoved the Manawatu forwards right off the ball. So powerful and so fast was their surge that Gilmer at halfback was taken by surprise and caught offside.

The referee's whistle blew. The Manawatu players' heads came out of the scrum hoping it signalled the end of the match. But their faces dropped when they realised it was for a penalty to North Auckland.

The tension around the ground was electric as Chippie Semenoff lined up for the kick. He ran forward and struck the ball towards the uprights. Those at the sides of the ground craned their necks to try and follow the course of the ball.

But those directly behind the posts knew exactly where it was going and could hardly bear to watch as the ball headed directly towards the left-hand post. As the ball struck it, which way would it fall? The crowd was hushed, the North Auckland players elated as it fell over the crossbar.

The final knockout blow. The Shield was heading north again.

58

THE SLIMMEST OF MARGINS

4 October 1978. Fraser Park, Timaru

Southern rivalry in promotion/relegation clash

South Canterbury and Marlborough are not two of the most glamorous unions. Both are based in rural areas and have to suffer the drift of their talent to the bigger unions. They were formed in the same year, 1888, and both were original members of the fledgling NZRFU.

This kick was further out and probably even more difficult for Marfell. Again he struck it well, again it headed towards the posts.

Both had experienced some glory days, though South Canterbury had some of their thunder stolen from their famous victory against France in 1961 by the grandmother with the flying fist, Hilda Madsen. Marlborough's year of glory was 1973 when they won the Ranfurly Shield and the Seddon Cup (contested between Marlborough, West Coast, Buller and Nelson). The following year they lost the Shield to…South Canterbury.

In 1977 Marlborough were still in the first division of the NPC, but they had a disastrous year, losing all their matches. Meanwhile South Canterbury had been cleaning up over the other teams in Division 2 South, winning all their matches and scoring 111 points against just 21 conceded. When they met in the promotion relegation clash at Blenheim, the demoralised Marlborough team was defeated by the fresh South Canterbury team 13-9.

The following year it was South Canterbury who had a torrid time in the first division, though they did manage to win two games. They didn't even come last, that dubious honour went to Hawke's Bay, but as they were the bottom South Island team that meant they were in the relegation play-offs again.

By contrast Marlborough had a great season in the lower level, winning all their matches and scoring 124 points against 33 conceded. Once again the two old rivals met in the bargain basement battle for survival. Who would win this time? Three thousand fans turned up in fine weather at Timaru to find out.

As expected, the match was tight and tense. At half time the scores were level at 3-3. Then after the break Marlborough centre Steve Marfell put the second division challengers ahead with a penalty kick.

South Canterbury fought back with a try to Neville Richards pushing them ahead by just one point. With ten minutes to go, they extended their lead to 10-6. Surely it would be enough in this dour contest. Then with five minutes of normal playing time remaining, Kevin Sutherland made a breakaway for Marlborough. He passed to winger Ian McAlpine who dived over to score. it was now 10-10.

As South Canterbury was the first division side that meant Marlborough had to win to go up. Marfell lined up the conversion attempt. It was a difficult kick. He struck it well. It sailed towards the posts, struck the crossbar and fell away.

Surely South Canterbury was safe now. But then with two minutes remaining on the clock, they gave away a penalty. This kick was further out and probably even more difficult for Marfell. Again he struck it well, again it headed towards the posts, but this time the ball fell short.

The South Canterbury fans breathed a sigh of relief. They had remained in Division 1 by the slimmest of all possible margins – nothing.

59

HADEN'S DIVE

11 November 1978. Cardiff Arms Park, Wales

McKechnie is the hero. Haden is the fall guy

The scene once again was that most hallowed piece of rugby turf – Cardiff Arms Park. While the battles between New Zealand and South Africa are always epic encounters, those between New Zealand and Wales seem to possess a mythical quality. They are never quite straightforward, always seeming to hinge on some piece of magic or skulduggery, either real or imagined. Because of the modern invention of the slow motion replay, the match that took place in 1978 was to become every bit as controversial as its infamous 1905 predecessor.

The Welsh had a strong team which was captained by J.P.R. Williams and featured the world famous Pontypool Front Row of Graham Price, Bobby Windsor and Tony "Charlie"" Faulkner. The All Blacks expected a closely contested match from a traditionally fierce foe.

That ferocity was displayed at the very first lineout where a flurry of punches resulted in a penalty attempt by Bryan Williams which fell short. A late tackle by the All Blacks enabled the Welsh to open the score with a penalty in the fourth minute of the match. Two minutes later the All Black fullback Currie had his jaw broken in a heavy tackle by Welsh Centre Steve Fenwick and Brian McKechnie came on as replacement.

The Welsh team was playing well and led 12-4 just before half time. Though the All Blacks brought the margin back to two points, the Welsh mounted extreme pressure late in the game and only a last minute tackle on scrum half Terry Holmes, one metre from the line, stopped them scoring and putting the match beyond doubt.

The All Blacks managed to get back into Welsh territory and with four minutes left to play, what was to become the most notorious lineout in the history of rugby formed 10 metres out from the Welsh 22.

As the ball was thrown in Andy Haden was seen to fall out of the lineout and the referee's whistle blew to award a penalty to the All Blacks. "You bloody cheats. You bloody cheats," the normally calm former Welsh player Barry John shouted, jumping up and down in the press box.

But the referee Mr Quittenton had awarded the penalty for a different incident,

PETER BUSH

A seemingly innocuous moment as Brian McKechnie takes the field as a substitute against Wales. High drama was soon to follow.

"You bloody cheats. You bloody cheats," the normally calm former Welsh player Barry John shouted, jumping up and down in the press box.

when Welsh lock Geoff Wheel had jumped off Frank Oliver's shoulder. Still that didn't stop Haden's Dive being endlessly replayed on TV with just as endless debate. Should the blatant attempt by Haden to cheat have negated Wheel's own attempt to cheat? Which was the graver offence of the two?

The answer depended on whether you spoke with the rounded vowels of the Welsh accent or the flat vowels of the Kiwi accent. By then it didn't really matter anyway because the match was over.

On the field, the fullback Brian McKechnie had stepped up to take the kick. The irony of the situation was that McKechnie had not been selected as a reserve until the morning of the match. He had not even been selected to make the tour originally and was not even a fullback, normally playing for his province Southland at first five-eighth.

At first the ball was not ready, so McKechnie tried to calm himself. It was a crucial kick for a man who had expected to watch the match from the stands.

"As soon as I hit it," he said later. "I knew from the way it felt on my boot that I had kicked the goal."

The All Blacks had won 13-12. It was a great kick under pressure, but the Welsh didn't care. All they talked about for the next day, week and year was Haden's Dive.

60

YOU'RE OFF – NO YOU'RE NOT!

1980. Stradey Park, Llanelli

Welsh and Kiwi Mafia invoke code of silence

It was not long before there was another tale to tell in the category of *What happened when the Taffies played the Kiwis.*

This time the All Blacks were touring Wales in 1980 to celebrate the centenary of the Welsh Rugby Union. Even though the All Blacks had been extended this prestigious invitation, the relationship between the two countries had soured somewhat on the previous tour. Apart from Haden's famous dive (sadly not nominated in the Academy Awards of that year) there was also the infamous raking incident of J.P.R. Williams by prop John Ashworth in the match against Bridgend.

In 1980, before coming to Wales, the All Blacks passed through the United States and Canada where the opposition were easily dealt with. But in Wales, in the course of two weeks, they were expected to play the four top Welsh clubs, Cardiff, Llanelli, Swansea and Newport, as well as the national side. Cardiff was a tough start which the All Blacks narrowly won 16-9. Then came Llanelli. In this match a very curious incident occurred.

Though Britain and New Zealand share the same language, the one word they can never quite agree on is rucking.

Though Britain and New Zealand share the same language, the one word they can never quite agree on is rucking. In New Zealand it is a method or using your feet to move a player on the ground out of the way when the ball is coming out. To the British it looks suspiciously like stomping.

During one of the forward encounters in the Llanelli match, All Black lock Graeme Higginson went into a ruck intent on getting the ball back for his side. A player was lying in the way, so he rucked him off the ball in the traditional New Zealand way.

The Scottish referee Alan Hosie did not see it that way and promptly ordered

Higginson off. In any other match he would have probably walked all the way to his early shower. But in the Llanelli side were two players of great pride. One was the Llanelli captain and one of the greatest fly-halves of all time, Phil Bennett. The other was centre Ray Gravell, a man passionate about both his rugby and his ancestry. Before Gravell made his debut for Wales, he worked himself up into such a frenzy in the changing room that he actually had to be calmed down before he could take the field.

Immediately, both foresaw that the following day's newspapers would say little about the rugby. Instead the headlines would trumpet the sending off in big bold lettering. They rushed forward to remonstrate with the referee, as did the All Black captain Graeme Mourie.

It quickly transpired that Higginson was being sent of for over rigorous rucking *of his own man*. Urgently, in words and gestures and with great passion from the Welsh side, the three players persuaded Hosie that it was inappropriate to send him off.

Higginson turned around and played on. Afterwards there was great interest in the incident, but those involved assured everyone that nothing had happened and it was all a "misunderstanding." The referee had merely indicated that the next time Higginson offended in such a way he would walk. The media pressed for answers, but the principal participants remained tight-lipped.

The All Blacks won the match 16-10 and went on to win the centenary test 23-3. The somewhat strained relations between the two countries were restored, helped by those urgent moments of quiet diplomacy on the field.

But some would say there were even greater ramifications than that. Somebody should have contacted the Pope to have the incident officially sanctioned as a miracle. It was the first time in the long history of rugby that a referee had been known to change his mind.

61

THE FLOURBOMB TEST

12 September 1981. Eden Park, Auckland

The most dramatic rugby match ever played

In 1981 the South Africans came to New Zealand and brought the country to a virtual state of civil war. Times had changed and for a new generation of New Zealanders rugby was no longer paramount. Far more important to many was the issue of apartheid in South Africa. While rugby took place behind barbed wire and shipping containers, outside pitched battles took place on the streets between police in riot helmets, armed with batons, and protestors similarly dressed in motorcycle helmets and wielding clubs.

In the test series meanwhile the old rivalry was just as intense as ever. New Zealand won the first test 14-9 in Christchurch, but lost the second 24-12 in Wellington. Yet again the final match in the series was crucial. Into this pressure-cooker came a new All Black cap, 21-year-old Gary Whetton. He hardly slept the night before as protestors kept up a chant around the hotel. The next morning the team left at 8.30 because police feared attempts would be made to block the Harbour Bridge.

"We walked through all the jeers from the protestors," Whetton said later. "It was quite an atmosphere. Once on the bus the police told us, "Right lads, just so you know. We are not stopping for anything – red lights, whatever."

"We formed this huge convoy on the motorway and cars would dodge in and out of traffic and stray between some of the vehicles in the convoy. But the police just moved through the convoy, got the cars out of the way...were just about getting their batons out at times. I can't describe the tension."

Eden Park was a fortress. The South Africans had been sleeping at the ground for three days. Time moved agonisingly slowly for the debutante.

"I think if I could have got out of the match at that stage, I would have. I was feeling the pressure that much," Whetton admitted.

On the relative cocoon of the rugby field, New Zealand went to a 16-3 half-time lead. In the second half the South Africans surged back while overhead a light plane swept over the ground dropping flour bombs. One felled All Black prop Gary Knight.

"Don't use water," referee Clive Norling shouted to the St. John's man. "He'll

"Don't use water," referee Clive Norling shouted to the St. John's man. "He'll turn to paste and we'll never be able to move him."

NEW ZEALAND HERALD

The match that had everything, the third test against the Springboks in 1981. The New Zealand players don't seem too concerned about Gary Knight being felled by a flour bomb.

turn to paste and we'll never be able to move him."

With one minute remaining on the ground clock, South Africa scored a try to level the scores 22-22. The conversion attempt was taken by Naas Botha, a master of timing and technique, dubbed "the man who never misses." The whole of New Zealand watched and waited as the well-struck ball spiralled upwards...and shaved the outside of the post.

Time was up on the ground clock, but because of all the stoppages to clear up flares, firebombs and flour bombs, the match continued.

Six minutes into overtime New Zealand won the match with a penalty kick, the spiralling ball succinctly symbolising the nation as it divided two poles. Two minutes later the referee's whistle ended the most dramatic rugby match ever played.

62

BAY GIVES WALLABIES PLENTY

1 September 1982. Rotorua Stadium, Rotorua

Never underestimate a hungry rugby team

In 1982 Australia toured New Zealand. They brought with them a formidable side that included the three Ella brothers Glen, Mark and Garry and a young wing called David Campese. They had held the Bledisloe Cup since 1979 and had just beaten the All Blacks 19-16 in the second test at Wellington when they travelled to Rotorua to take on Bay of Plenty in the midweek fixture.

The result of the match appeared to be a foregone conclusion. Facing the Aussies was a Bay side that contained two All Blacks in Hika Reid and Frank Shelford but which had lost all five of its matches in division one that season. What was in doubt was not whether Australia would win, but by how much. Bay of Plenty were clearly just an appetiser in the tourists' run up to the final and deciding test in Auckland.

Unfortunately for the Aussies, nobody informed the Bay players of the minor role they were supposed to adopt that day. Least of all the Bay hooker Hika Reid who had a storming game, raising his level of performance to a plateau that inspired the rest of the team. Even the backs, who had been unimpressive all season, were able to lift themselves to match an Australian backline that included Gary Ella and David Campese (who came on as a substitute).

The home team went ahead in the second minute with a penalty goal to Preston. They scored their first try when their speedy winger Basham won the race to a kick ahead by Cameron. The Aussies tried to hit back but Cameron counter-attacked from deep inside Bay territory to run all the way to the try line with Australian captain Slack in hot pursuit.

Unfortunately for the Aussies, nobody informed the Bay players of the minor role they were supposed to adopt that day.

Just after the start of the second half Preston scored from a midfield break. Then McKillop launched a high kick in the direction of the Australian posts. It was the mighty Hika Reid who gathered up the ball and dashed over for the try. He got his second try soon afterwards when another high kick by Cameron was fielded by Glen Ella who was caught in possession. The ball was ripped off him by the Bay forwards who were following up and it was Hika the Lion who scored his second try.

With minutes in the match remaining, Bay of Plenty had an attacking scrum five metres from the Australian line. McCarroll ran wide with the ball then inpassed to the flanker Gordon Tietjens (later to become the New Zealand sevens coach) and he crashed over to score near the corner. The Bay fans' day was made complete when Preston kicked a difficult conversion.

The 40-16 victory delighted the 10,000 local spectators and the Wallabies had the grace to acknowledge this amazing victory by clapping the Bay of Plenty team off the field.

It was something to savour, the wholesaling drubbing of a national side by a team that would finish the season squarely at the bottom of the first division table.

63

BUTCHER'S BOY FED TO LIONS

18 May 1983. Eden Park, Auckland

The story of how J.K. made the grade

It is every young boy's dream to be plucked from obscurity and pitched straight into the cauldron of a big match. It has happened, but very rarely and there is usually a bit more to the legend than meets the eye. The story goes that John Kirwan was taken by the scruff of his neck out of third grade rugby by John Hart and tossed straight into battle with the mighty British Lions. It was almost like that.

In 1983 Kirwan worked in his butcher's shop, drove around with a surfboard strapped to the roof of his car and was playing third grade rugby for Auckland. He was a strapping lad standing 1.90m tall and weighing about in at about 90kg. His style was a mix of a bristlingly healthy aggression and blistering speed.

Kirwan's rise to instant stardom actually started the year before in 1982. Auckland coach John Hart got a tip-off about Kirwan from a friend, former All Black centre Paul Little (who had played his club rugby for Marist, Kirwan's club). One Saturday, on his way to Eden Park, Hart had stood outside the fence at the Marist ground and watched this "big, raw boned winger."

Then at the start of the following season Kirwan got a call from another friend of Hart's, Peter Murdoch, inviting him to play for the Barbarians. Kirwan refused. He had already planned a trip to Australia with his friend. When he got back, the

Baa Baas still wanted him to play.

He played and scored a try in the 7-3 victory. By now Hart had seen enough and selected Kirwan for the Centennial team that was to play against the President's XV to celebrate Auckland's centenary. First though, Hart had to get permission from Kirwan's parents. Mr Kirwan was eager, Mrs Kirwan not so, but she was finally persuaded.

The roar was so loud as he ran onto the field that he said it was like having stereo speakers pressed against his ears.

Hart realised the pressure that would fall onto Kirwan and told the media that Kirwan had been chosen for the centennial game to represent the youth of Auckland. Kirwan was nervous before the match, but Hart told him just to concentrate on playing the basics of his position. Unknown to Kirwan, he also told the rest of the team to look after the baby of the team. Kirwan played well and just missed out on a try when his Marist teammate Kevin Boyle beat him a loose ball to score. Kirwan still believes that Boyle won the race because he came from an offside position.

After an NPC match against Waikato, suddenly Kirwan found himself playing against the British Lions in what for them was virtually fourth test of their tour. Auckland had not beaten the Lions since 1930 and in the last fixture in 1977 had performed poorly, losing 34-15. It was Kirwan's third first grade match and his first big game. The roar was so loud as he ran onto the field that he said it was like having stereo speakers pressed against his ears.

Grant Fox kicked the first points, but the Lions responded with a dropped goal. Then the Lions scrum half Terry Holmes made a break to feed Irwin who scored. A penalty apiece meant the Lions lead 12-6 at the break.

Auckland upped the pressure in the second half. Mills picked up the ball after a kick ahead by Stanley. He beat one tackle and then passed to former All Black wing Gary Cunningham who scored in the corner. With just four minutes left in the match Fox put over the drop goal that put Auckland ahead by one point for a 13-12 victory.

But the match could have been lost earlier when the English were on the attack and kicked ahead. The ball rolled ahead over the Auckland line with the English centre Clive Woodward bearing down on it. It was the 18-year-old "raw boned winger" who sped across to beat him to it and save a certain try.

64

LIONS DEFANGED

16 July 1983. Eden Park, Auckland

Stu Wilson breaks try-scoring record

After the defeat by Auckland in just the second match of the tour the Lions did not lose again to a provincial side except for a narrow 22-20 loss to the Ranfurly Shield holders Canterbury. They had managed some impressive victories against some of the lesser unions (beating Southland 41-3) but an international team is judged in test matches and here they were woefully inadequate. By the time the Lions limped into Eden Park for the final match of the series they were already 3-0 down.

They faced an experienced New Zealand team. So experienced in fact that the front row of Gary Knight, Andy Dalton and John Ashworth was known as "the geriatrics." Experience told the All Blacks that they were in for a hard match because the Lions would want to avoid a whitewash and salvage some pride from the tour.

That was one scenario. Another was that the Lions would adopt the attitude of their manager Willie John McBride who had been dubbed "Whinging Willie" for his constant complaints about referees. So would they play like Lions or like lambs?

The All Blacks were ahead after just three minutes when Hewson kicked a penalty. Six minutes later he got another one. Then Pokere slashed 40 metres upfield through the Lions meek defence to be stopped just short of the line. From the scrum Hobbs went over for his first try for New Zealand. An overlap created by Fraser coming into the backline brought a try to Wilson – his 17th in tests (beating Ian Kirkpatrick's previous record of 16).

The All Blacks were playing with awesome efficiency and penetration and the Lions were lucky to be down by just 16-3 at half time. They allowed the tourists only two minutes respite after the restart before Wilson scored his 18th test match try. Five minutes later the All Blacks threw up a high kick. The Lions' fullback Evans caught it, but Hewson was following up. He wrestled the ball from the Lion's paws and scored.

With the British team reeling, the All Blacks went into overdrive. Dunn charged right up to the Lions line where he was held. The ball came back to Haden from the maul and he forced his way through under the posts. When the Lions tried a desperate counter-attack from inside their own 22, MacNeill's long pass was

NEW ZEALAND HERALD

The two opposing hookers and captains Andy Dalton and Ciaran Fitzgerald. Behind them the fans invade Eden Park to celebrate New Zealand's 4-0 whitewash.

intercepted by Wilson who went over for his 19th test match try.

With 15 minutes to go and a score of 50 or more in prospect, the All Blacks felt sorry for the hapless Lions and eased off the throttle. The final score was 38-6 (a pair of meagre penalty goals to the Lions) and the defanged and declawed Lions limped back to the British Isles humbled by one of the most impressive and dominating displays ever seen by an All Black team.

The All Blacks were playing with awesome efficiency and penetration...

65

THE CHALLENGE OF CHALLENGES

14 September 1985. Lancaster Park, Christchurch

The greatest Shield match ever?

Until the 1980's the Auckland team of 1960-63 held the record for the number of Ranfurly Shield challenges with 26. By 1985 a committed Canterbury team coached by Grizz Wyllie had run up 25 consecutive defences.

Sportingly, Canterbury set up the ultimate match by ensuring the 26th challenge was from Auckland, giving their northern rivals the opportunity to not only try and prise the Shield off them, but to retain their record.

The match took place on a sunny Canterbury day, with a huge crowd spilling onto the field. Auckland coach John Hart had gone in for a low-key build up and his relaxed team took the tense holders by surprise, playing sublime rugby.

Almost straight from the start a high kick was botched by Canterbury and Joe Stanley scored a try. Then selfish running by Canterbury back Vic Simpson inside his own 22 gifted a try to John Kirwan. Before the half-time break Auckland prop John Drake smashed over for another. The huge Canterbury crowd was silent. Auckland was annihilating their beloved team 24-0.

At half time Alex Wyllie tossed the match ball to his beleaguered players and told them to get hold of it – because they couldn't play without it. Straight after the break they came back with halfback Bruce Deans sprinting down the touchline for a try. But then Auckland responded with a try to McDowell. That took the wind out of Canterbury's sails. Surely they were beaten now.

The huge Canterbury crowd was silent. Auckland was annihilating their beloved team...

But that was when Canterbury hearts really started pumping and they made a scintillating comeback. Winger Craig Green made a dash down the field to score in the corner. Robbie Deans converted from the sideline. Then Deans goaled a penalty. Still 15 points behind they mounted a series of raids on the Auckland line. Wayne Smith dived over for a try. Then lock Albert Anderson caught a high bomb and ran straight through the tackle of Grant Fox like a charging bull to score.

Canterbury was on fire. Their fans went wild.

And now they were just five points behind. But the seconds were ticking away. There was time for one last attack. Wayne Smith hoisted a high ball which came down out of the clouds across the Auckland line. It must have felt as if the whole Canterbury team was thundering after it as John Kirwan desperately tried to catch it. The ball bobbled off his fingertips and bounced away over the dead ball line.

At that moment the final whistle went. Auckland had won 28-23, winning the Shield and holding onto the record by the thickness of John Kirwan's fingertips.

66

PANTHER STRIKES

18 September 1985. Fraser Park, Timaru

Michael Jones makes an awesome debut

When Michael Jones left school and joined the Waitemata third grade side it wasn't long before he was promoted into the senior side. At first he found it strange playing with guys who had been his heroes. Even stranger was picking up a copy of the Saturday sports paper, seeing a photograph of a player jumping for a lineout and realising it was him.

In 1984 he was selected for Auckland under 18s and the following year made the Auckland Colts. They travelled down to Lancaster Park to play the curtain raiser for the "Challenge of Challenges." Jones's team lost, but he was thrilled to watch the Shield game and celebrate Auckland's stunning victory. After the match he went to visit his grandmother who lived in Christchurch. While he was there he got an unexpected phone call.

Unknown to him, before the game, the Auckland coach John Hart had decided that if Auckland won the Shield, he would blood two of the Colts in the A team for a tour match against South Canterbury in Timaru before the team returned to Auckland. One of the players was Bernie McCahill and the other, Hart told the stunned colt on his auntie's phone, was Michael Niko Jones.

Jones hurried away to join the team for the journey to Timaru. On the bus he sat next to fellow Samoan Peter Fatialofa but he was so awestruck at being on the same team as him that he did not say a word to him. In fact he did not talk to anybody for the whole journey.

Before the game Hart told him, "Do your best and don't try and do too much."

"All I remember is the thrill of playing the game," Jones said afterwards. "I am sure I was nervous but it was outweighed by the excitement and keenness to get out there and play with these guys."

"This panther-like creature running around with simply boundless energy and an amazing range of skills..."

What the 20-year-old Michael Jones was able to bring to the team was a level of athleticism never seen before. Grant Fox was impressed, "This panther-like creature running around with simply boundless energy and an amazing range of skills," was how he described Jones that day. It was a great debut. Jones played like a whirlwind and scored three tries.

Yet he was mightily embarrassed when the team returned to Auckland. Jones was included in the ticker-tape parade down Queen Street for Auckland winning the Shield – even though he had not played in the Canterbury match.

67

NEW TEAM WITH OLD FACES

18 September 1985. Trafalgar Park, Nelson

North Harbour win Division 3 at first attempt

In 1985 Auckland was getting too big for it's rugby-playing boots. The city accounted for about one third of the population of New Zealand and was the most powerful union, holding the Ranfurly Shield and regularly cleaning up in the NPC. In 1926 the South Auckland Union had been formed (changing its name to Counties in 1956 and to Counties Manukau in 1996). By 1985 it seemed about time for a northern split with a lot of good rugby players north of the Harbour Bridge who were struggling to get into the Auckland "A" side.

After much discussion North Harbour was formed in 1985 as the 27th and youngest union in New Zealand. (Though the region is locally known as the North Shore, a club side had already taken that name.) There were some who had played their club rugby on the Shore, such as Terry Wright, who decided to stay with Auckland. Others decided to take a chance on the fledgling union.

One was winger Gary Cunningham who had played five tests for the All Blacks between 1979-80. Another was the new North Harbour captain Wayne "Buck" Shelford, who was shortly to be selected for the All Blacks, and Frano Botica who would become an All Black the following year. Even so, as the new boys on the block, North Harbour had to start right from the bottom.

The previous season there had been a Second Division North and a Second Division South. In 1985 the format was changed to form a third division, but with teams divided into North and South Island sections. North Harbour was placed in the North Island section along with Poverty Bay, East Coast and Thames Valley.

North Harbour played their inaugural NPC match at Takapuna in mid-September against Thames Valley. The babies of the NPC got off to a nervous start, but the superior playing skills of the city slickers soon got the better of their country cousins as they ran in a hatful of tries, including three to winger Wayne Hill in the 39-3 win.

Next they were asked to play two games in five days. First they went down to Poverty Bay where their opponents lived up to their name by getting no points on the board. North Harbour scored 35 in an emphatic victory which saw Paul Feeny run in four tries.

Six days later in Ruatoria the Harbour machine ran in six tries against the East Coast, running the ball at will in a 37-4 win. In their first three games Harbour had scored over a century of points (111) against just 7 in reply. That gave them the right to play against the South section winners Nelson Bays to see who would be promoted.

It was the mismatch of the decade. North Harbour already had the player power to seriously beat most second division sides. At first Nelson Bays did manage to make a fight of it, holding the city boys to a lead of just 15-7 at half time.

In the second half though their players could do little more than direct traffic as Harbour ran in eight tries. The former All Black wing Gary Cunningham got a hat trick, Buck Shelford got two and Frano Botica kicked 20 points to sweep Nelson Bays aside 54-10 and win promotion to the second division at North Harbour's first attempt.

All of the other third division sides wished them "bon voyage". While many of those players now had the opportunity to tell their grandchildren they had opposed the great Buck Shelford, none of them fancied another season of trying to stop his rampaging runs.

...as the new boys on the block, North Harbour had to start right from the bottom.

68

THE NOT SO CAVALIER CAVALIERS

June 1985 – June 1986. New Zealand High Court – Ellis Park, Johannesburg

When Apartheid finally became bigger than rugby

In 1985 the All Blacks were invited to tour South Africa. After the virtual civil war that had erupted in New Zealand in 1981 there was a lot of dispute about whether or not the invitation should be accepted. Prime Minister David Lange informed Ces Blazey, the Chairman of the NZRFU, that he was against the tour. However, at the meeting of the unions, despite two of the biggest unions Auckland and North Harbour casting their vote against the tour, it was officially sanctioned.

That was when a group of lawyers decided to employ the legal system to stop the tour. They used two rugby club members who were also lawyers, Phil Recordon and Paddy Finnigan, to bring suit against the NZRFU that the tour was not in the best interests of New Zealand rugby. In the High Court the Chief Justice Sir Ronald Davidson rejected the claim, but the Court of Appeal reversed his decision.

The court action began six days before the All Blacks were due to leave, but the lawyers then applied for an injunction to ensure the matter had been heard by the courts before the tour went ahead. That was impossible, so the tour was off.

While expressing relief at this decision Lange threw the cat back among the pigeons by saying the All Blacks were entitled to go as individuals. The All Black players then regrouped at the home of former All Black Doc Cunningham to plan a rebel tour. They were all for it except John Kirwan, Grant Fox and David Kirk.

However the tour did not go ahead because legal advice was given that Andy Dalton, Brian Lochore and Andy Haden would be in contempt of court if they went, and the other players did not want to go without them.

As "D DAY" approached the players slipped out of New Zealand in small groups.

The following year some of the players heard whispers that the unofficial tour might be on again and plenty of cloak and dagger arranging went on behind the scenes. When told of this Ces Blazey was at a rugby congress in Oxfordshire. He immediately sought out Danie Craven, Chairman of the South African Rugby Board. Craven was evasive. It would later transpire that he had been secretly making arrangements with the New Zealand players without informing the New Zealand Union.

As "D DAY" approached the players slipped out of New Zealand in small groups. Many of them were already scheduled to play in the IRB centenary matches in Britain and then flew straight on to South Africa. In the end Grant Fox decided to make the tour. John Kirwan stuck to his guns and didn't, while David Kirk pulled out at the last minute.

In South Africa the rebel side was known as the "All Blacks" but in New Zealand and world-wide it carried the more derogatory nickname of "Cavaliers."

Disaster struck the Cavaliers in only the second match when captain Andy Dalton's jaw was broken by Northern Transvaal player Burger Geldenhuys.

The Cavaliers led 15-12 in the first "test" with minutes to go but then Naas Botha kicked a penalty and Bok winger Carel Du Plessis won the race to the ball to score in the final minutes. They won the second "test" literally by a fingertip. In that game the Cavaliers went out to an early lead but then pulled back to just one point behind at 19-18. The ball came back to Botha who steadied himself for the drop goal. Jock Hobbs came haring through and just got a fingertip to the ball deflecting it off course and away from the posts. John Ashworth caught it and gave it a mighty boot. The referee's whistle blew as the ball soared deep into the crowd.

That squared the series but the Cavaliers were blown away 33-18 in the third "test" despite leading 12-9 at half-time. The final "test" was lost 24-10 at Ellis Park.

All the Springbok players were awarded full test caps for their involvement. In contrast, all the Cavaliers were banned from international rugby for two matches by the New Zealand union.

No doubt if they had won the "series" in South Africa however, the tour would have been hastily upgraded to full test status.

69

THE BABY BLACKS

28 June 1986. Lancaster Park, Christchurch

New Zealand's third team takes on the French

With all the Cavaliers banned for two matches the selectors had some problems when the French came to visit. France was on a post-season (Northern Hemisphere season) round the world tour which took in Argentina and Australia, as well as New Zealand. They had visited these shores as recently as 1984 losing both the tests, though the All Blacks had only won the first test by one point. On the 1986 tour they played just one other match in New Zealand, beating North Auckland 30-9 before the test match.

The All Black selectors were able to choose only four players with test match experience in Kirwan, Stone, Kirk and McGrattan. This created the unusual situation of a large number of new caps being selected. Because many of them were relatively young the press soon started dubbing them the "Baby Blacks."

The nine Baby Blacks originally selected were Greg Cooper, Joe Stanley, Terry Wright, Frano Botica, Mike Brewer, Mark Brooke-Cowden, Andy Earl, Gordon Macpherson, Brett Harvey and Kevin Boroevich. This was Boroevich's first test, though he had played in 13 matches on tour for the All Blacks over the last couple of years. So technically he was more accurately described as a "Toddler Black."

Many people were surprised that the regular Auckland hooker Iain Abercrombie was not even chosen for the squad. Instead on the reserve's bench was his provincial understudy, a young man by the name of Sean Fitzpatrick. During the week leading up to the test, selected test hooker Bruce Hemara had to withdraw with a rib injury and so Sean Fitzpatrick became the 10th baby in the team. He was 23 years old.

The weather conditions were good on the day of the match, but a crowd of only 24,000 turned up to watch. Perhaps they were missing the regular stars or perhaps they expected the Babies to have their bottoms smacked by the much more experienced French team.

It certainly seemed that way. The French forwards were heavier and Fitzpatrick found the scrums really tough. He could hardly breathe in them. "Every scrum was an absolute nightmare to be in," he said later.

Cooper put New Zealand ahead with a drop goal then Botica made a break down the blind side and kicked ahead. Kirwan gathered the ball and ran on to the French 22 where he passed on to Harvey who put Brewer over for the try. The French were unable to score a try during the game and with ten minutes to go points from penalties meant the Baby Blacks were leading 18-9.

As the time ticked away, eager for his first test win, Fitzpatrick kept thinking. "How long? How long?" When the whistle finally blew he felt an immense relief. The Babies had come of age.

70

THE BATTLE OF NANTES

15 November 1986. La Beaujoire Stadium, Nantes

What did the French have in their water on that day?

At the end of 1986 the All Blacks went on a short eight-match tour of France. Coming into the second test they had won all their matches including the first test the week before by 19-7. The French selectors had made several changes including bringing in the rugged Pascal Ondarts at prop. The 31-year-old was making his international debut and he was to make a major impression on the game.

Nobody knows exactly what was put into the French water just before the game. Some said it was wine, others a hint of French passion, while there were those who said it was something that would have failed a drug test. Years later Buck Shelford was to get into trouble by making that allegation. But typical of the man he stood by his words.

"I will always believe they were on something that day. I know the type of commitment and excellence it takes to go 80 minutes the way they did in Nantes. It's just not possible to keep up that sort of aggression for that long without having a flat period."

Shelford should know. Former All Black coach Grizz Wyllie said of him, "The more sprig marks he has over him, the happier he seems to be. The harder the action is, the more it appears to spur him on."

The match was only Shelford's second test, though earlier that year he had toured with the Cavaliers and hardened up on the big men of South African rugby.

But even that did little to prepare him for the day in France when the French team came out of the tunnel like out-of-control combine harvesters. Straight from the kick-off it was all fire and fury from the French. "This is going to be one helluva rugged test match," Shelford thought to himself as the battle began without any of the benefits of the Geneva Convention.

...the French team came out of the tunnel like out-of-control combine harvesters.

In the next ruck he received a "Bonjour Monsieur Shelford" when a Gallic boot careered straight into his head. Shelford rose groggily to his feet, felt in his mouth and started counting. He realised it was "au revoir" to three of his teeth as he ran upfield to follow the play.

He was soon in the action again, snatching up a loose ball and turning to set up the maul. The French hooker and captain Daniel Dubroca came through and finding the ball well protected, he kicked straight between Shelford's legs trying to dislodge it. Instead he connected with another type of round object and Shelford winced with pain. But he played on.

It was the 15th minute of the second half before he was finally bludgeoned off the field, when his head was the target of a full-blooded charge by the French prop Garuet. This time Shelford was down and out. Back in the dressing room the medical staff carried out an inventory of the damage.

Not only did New Zealand lose the test 16-3, but they also found Shelford was missing one of his testicles. It had gone "absent without leave" as a result of the kick by Dubroca. It was returned to its rightful position and the scrotum sewn up again with 18 stitches. As Shelford had predicted it had been hell of a match.

But then he was hell of a hard player. One who was prepared to lay it on the line and give everything for his teammates and his country – including his manhood.

71

J.K.'S TRY

22 May 1987. Eden Park, Auckland

John Kirwan scores a spectacular solo try

For years the administrators had dithered, but finally in 1987 the first Rugby World Cup took place in Australia and New Zealand. The sixteen best teams from around the world were invited to play, except for South Africa who, because of apartheid, were the pariahs of world rugby.

The New Zealand squad chosen contained a mix of the Cavaliers and the Baby Blacks from the previous season, but coach Brian Lochore instilled team spirit early when he told them, "The only way we are going to win this cup is by getting together and forgetting what happened last year."

The team was chosen to play a style of game where the ball was moved wide to big wingers who could stand in the tackle and offload to support players backing up quickly. Two of the key men in this tactic were winger John Kirwan and his Auckland team-mate, the dynamic young loosie Michael Jones.

The tournament got under way on 22 May when New Zealand kicked off against Italy. The match was preceded by a torch bearing ceremony by former All Black-great Waka Nathan.

No one was sure what to expect from the match. New Zealand had never before played Italy who had twice been runners up in the FIRA championship. The first points of the match went to New Zealand scored by *nobody*. The referee adjudged that when Innocenti the Italian captain tackled Buck Shelford who had the ball at his feet, he had stopped a certain try. He awarded a penalty try.

The first points of the match went to New Zealand scored by *nobody*.

After that the game opened up as the inexperienced Italians folded under the All Black pressure. The All Blacks scored a total of 10 tries in the record 70-6 win. The most memorable one was a 90-metre run by John Kirwan.

Late in the match after New Zealand had scored another try through McDowell, the Italians restarted long. The ball was fielded by Kirk who passed to Fox, who sent the ball on to Kirwan who was standing deep inside his own 22.

At the beginning of the season Kirwan had decided he wanted to be able to score tries from any part of the field. He trained accordingly, building up his strength and stamina, but ensuring it was not at the expense of speed. He also improved his stepping by practising with cones.

When he received the ball against Italy it was a situation he had already imagined long before. The Italians were mesmerised as he weaved in and out of virtually the whole team. The only person who could keep up with him was his shadow Michael Jones who could see the cuts coming before J.K. made them from playing with him for Auckland.

But Jones soon stopped when he saw Kirwan was going all the way to the try line to score the most thrilling try of the tournament. J.K. received a standing ovation from the crowd as he jogged back.

72

WORLD CHAMPIONS

20 June 1987. Eden Park, Auckland

The All Blacks win the inaugural Rugby World Cup

On 20 June 1987 the whole nation was asking one question – could the All Blacks win the very first Rugby World Cup?

NEW ZEALAND HERALD

Eden Park 1987. It's official – New Zealand is the champion of the world.

They were playing at home and their closest match so far during the tournament (against Scotland) had seen them win by a clear 27 points, but they were also playing against France. The French are never easy opponents. Some days they play without spirit or discipline. Other days they are brilliant and can beat anybody. They had shown that just a week before in the semi-final against Australia.

The Australians had believed it was their right to contest the final with New Zealand but the French had won an epic match with a last minute try by Serge Blanco. How good would the French be in the final? No one knew.

Before the match could start at Eden Park, a rooster painted in French colours had to be removed from the field. Immediately the All Blacks went on the attack and finally were rewarded with three points from a drop goal by Grant Fox. When he tried another drop goal a little later, the ball was deflected off the French first five-eighth Mesnel and Michael Jones was able to pick it up and score. By half time the All Blacks led 9-0. However, many wondered if this was enough because in the second half they faced the breeze and a low sun.

Before the match could start at Eden Park, a rooster painted in French colours had to be removed from the field

Those fears seemed justified when France scored first after the restart with a penalty by Camberabero. Then New Zealand drew further ahead with two penalties by Fox. Yet the French still could have come back from there. A try was needed to break their resistance.

The All Blacks drove up to the French 22 and Kirk passed to Fox who off-loaded to Jones who made ground before he was held up. Kirk took the ball up again and sprinted between Sella and Blanco to score.

It was Kirk again who made a break from the kick-off. He sprinted 50 metres up the field before he was tackled and the ball spilled out of his grasp. Buck Shelford was following up. He snatched up the ball and sent out a quick pass to John Kirwan who scored in the corner.

The French rallied to score a consolation try two minutes from the end, but it was too late. New Zealand had won. David Kirk became the first person to hold aloft the Webb Ellis Trophy. There would be many other World Cups, but there could only ever be a first one and New Zealand had won it.

73

UNWANTED MAN SAVES THE DAY

17 September 1988. Eden Park, Auckland

Auckland comes within one metre of losing the Shield

Frank Bunce is famous as the man Auckland didn't want, but the All Blacks did. He spent what should have been the prime of his playing career trying to get into the Auckland side, before finally having to go to North Harbour to get regular rugby and finally into the All Blacks. But not a lot of people know that if it weren't for Bunce, Auckland would never have achieved their record-breaking run of Shield defences.

Bunce was first selected for the Auckland C team in 1983 on a short tour up north. The following year he made it into the Auckland B team, scoring two tries on his debut against King Country. In 1985 he was selected to the A squad for a game in Rotorua, but didn't get on the paddock.

In 1986 he got the opportunity to trial for the Baby Blacks but did not perform well as he was suffering from an ankle injury. Joe Stanley got the nod ahead of him and was also ahead of him for the number 13 jersey in the Auckland side.

It was only when the All Blacks went to Australia that Bunce finally got to start an A game for Auckland. He got a try in the 82-6 Ranfurly Shield win over Horowhenua.

In 1987 new Auckland coaches Maurice Trapp and Bryan Williams still preferred Smoking Joe, and even newcomer Craig Innes, to Bunce. In 1988 Bunce managed eight games for Auckland. Only one was in his preferred position of centre. In that match against Manawatu, Bunce scored three tries and was made man of the match. He played three matches on the wing and in four came on as a replacement. It was in one of these, almost the last match of the season against Otago, that he saved the Shield for Auckland.

As the move unfolded Bunce could see exactly what was going to happen.

Bunce went on to replace the injured Terry Wright on the wing. At 13-9 Otago were winning a surprising amount of lineout ball through Brewer and Knight and succeeded in keeping Auckland pinned in their own territory for long periods. Then Auckland counter-attacked, bringing the fullback Lindsay Harris into the play. On the wing, Bunce could see the move unfolding and had the instinct to move across and cover for the fullback, but he decided not to. A moment later the Otago wing Noel Pilcher came screaming through for the intercept and sprinted to the line

NEW ZEALAND HERALD

Frank Bunce (left) does it with his eyes shut here, but he needed them wide open to save the Ranfurly Shield for Auckland with a last-ditch tackle.

directly through the gap Bunce would have plugged. Pilcher dived theatrically over the line to score.

After the restart Otago went straight onto the offensive again. They attacked off a scrum, drawing the Auckland backs into the tackle. As the move unfolded Bunce could see exactly what was going to happen. Pilcher scorched in from the blindside wing, beat three tacklers and headed for the line.

A metre short of it he was met by Bunce in a textbook tackle that sent him to the dirt and saved a certain try. At that stage the score was 17-12. But for the tackle, with conversion it probably would have been 23-12 and there would have been no way back. Instead Auckland clawed their way to victory and retained the Shield.

When Bunce went into the Auckland supporter's club afterwards someone shouted at him, "Good on you Frank, you saved the Shield for us today."

But it wasn't until the Monday morning when he saw the photograph of the incident in the *New Zealand Herald* that he realised for the first time just how close Pilcher had been to scoring.

74

MATCH ABANDONED

1988 Tucuman, Argentina

The Maoris play inside a cage in Argentina

In 1988 the Maoris made a tour of Italy, France, Spain and Argentina. The tour was very successful with the Maoris winning ten of their 12 matches, losing one and drawing one, but it had many unusual aspects.

In the first game against the Italian Barbarians at L'Aquila the opposition included such "locals" as South African Naas Botha, Welshman Robert Jones and even former All Black Craig Green. That was won 57-9, but the next game against Toulon in France was a nightmare. The violence started from the kick-off and carried on through most of the match, though it was one of the Maoris, prop Arthur McLean, who was sent off for punching. That match was won 22-9.

In Rodez against the Southern French Selection there was trouble at scrum time when the opposition kept charging over the mark. Unfortunately the referee did not speak English or Maori and the Maori captain Buck Shelford did not speak French so he led his team to the side of the field to summon the interpreter. Straight-talking Shelford told him in no uncertain terms to tell the ref that unless the scrums were fixed the Maoris were "off."

The referee got the message and there was no further trouble. Until the next day. Then the story submitted by the New Zealand Press Association reporter was altered in New Zealand to make it appear the team had really walked off.

When the Maoris beat the French Barbarians the captain of the opposing team was another former All Black – David Kirk. After France they travelled on to Argentina which is one of the hardest countries to tour. They speak a different language, there are long distances to travel and the welcome is sometimes inhospitable both on and off the field.

The main match of the tour was against Tucuman. Two years before Tucuman had fought out an 18-all draw with the touring French team. But the crowd behaviour had been so bad the team had been banned for two years. As the Maoris ran onto the field in the 40-degree heat they saw that the hostile-looking crowd was kept behind wire cages.

The matches in France were mild compared to what happened in Tucuman. The local players did not appreciate good old-fashioned New Zealand rucking and came up fighting and punching. More trouble came when halfback Brett Iti was downed with a high tackle by the opposing first five-eighth. Prop Steve McDowell manhandled the ref while complaining about this incident and was promptly sent off, despite the entreaties of Maori captain Buck Shelford and the Tucuman captain to keep him on the field.

The worse the violence got on the field, the worse it got behind the cages. Anything the crowd could get their hands on including beer bottles came raining over the fences. At one point the Maoris huddled in the centre of the field to escape the barrage while the Tucuman players went and pleaded with the fans. They didn't want to get banned for another two years.

Finally it got so bad that with five minutes to go and the Maoris leading 12-3, the game was called off. Jasin Goldsmith was hit in the face by a bottle as he walked off. The Maoris' Argentinian liaison officer had to be carried off after a bottle hit him on the knee. And a fond farewell to Argentina was administered to Buck Shelford who was whacked over the head with an umbrella.

The worse the violence got on the field, the worse it got behind the cages.

75

LIGHTNING STRIKES

1989. Lancaster Park, Christchurch

Did Terry Wright score the fastest try ever?

The scene was set for another epic contest between the two old rivals Auckland and Canterbury. On a sunny day Lancaster Park was brimming with an eager crowd waiting for the match to start. In the stands below the Auckland coaching staff gave final instructions to their troops and then made their way towards the seats.

The players ran onto the field and lined up for the kick-off. The referee made sure the two teams were ready and blew his whistle. The ball was kicked off. It came through to the Auckland centre Craig Innes who instead of running the ball back as he was expected to do, stabbed a speculative kick into the Canterbury half.

There fullback Robbie Deans had still not appreciated the match had started. Suddenly he saw the ball coming towards him and watched as it flew over his head. Following up fast was the Auckland winger Terry Wright.

Wright had first come to prominence years before when he went on a tour with the Auckland C team in 1983. The other C players had quickly realised that this fragile seeming stick-and-bone man possessed exceptional pace. Their tactic for the entire tour had been to get him into a position where he could use it.

In his subsequent rugby career Wright made a mockery of those who thought he would be blown over in a strong wind. He played in over 100 matches for Auckland and went on to play score 19 tries for New Zealand in just 30 internationals – a strike rate that was quite incredible until the arrival of the even more incredible Christian Cullen years later.

With his blistering speed Wright was the last man the one-eyed Cantabrian fans at Lancaster Park wanted to see with the ball in hand...

With his blistering speed Wright was the last man the one-eyed Cantabrian fans at Lancaster Park wanted to see with the ball in hand and the try-line beckoning. They began to choke on their hot dogs, their early enthusiastic cries of encouragement dying in their throats as he took off. In less than five seconds with the cover defence piling after him, Wright sped across the Lancaster Park turf and touched down.

The scoreboard man had not even had time to oil his rollers, the Auckland coaching staff had not even got to their seats. Up in the gantry the commentators looked at each other, scratching their heads and comparing the little hands on their watches. Wright had scored after just 17 seconds!

The Canterbury team trouped back in a state of shock to line up for the conversion. They shook their heads and sent injured glances at the referee. "Yes, we know we said were ready to start, but we weren't *actually* ready to start."

Canterbury fought hard throughout the match to make up those first few seconds of inattention. But they never did. In the end the try that Wright had scored while Canterbury were still wiping the sleepydust out of their eyes was what really separated the two teams in Auckland's victory.

76

THE END OF AN ERA

18 August 1990. Athletic Park, Wellington

An unbeaten run of four years comes to an end

On the evening of 18 August 1990 even hardened New Zealanders were known to shed a tear or two and go rummaging though the old trunk in the garage to pull out their black armband. It was a day to rank with the end of the two World Wars, a day that was filled both with pride and yet a deep sense of loss. It was the end of an era.

Almost half a decade before, on the blood-soaked field at Nantes in France, the All Blacks had fallen under the onslaught of a French storm. Since that dark 16-3 loss the All Blacks had proved invincible.

The long run of victories started in 1987 when the team won the first World Cup. They were a freakishly good side, led first by the little general David Kirk and later by the intimidating Buck Shelford. In the middle of the front row was a young and aggressive hooker Sean Fitzpatrick. Either side of him were John Drake, who many considered to be the best prop in the world, and judo black belt Steve McDowell. Also in the pack were the Whetton brothers and the Iceman Michael Jones.

In the backs was the kicking machine Grant Fox, "smoking" Joe Stanley, John Kirwan (still today the All Black's leading tryscorer with 67 in all matches), and at fullback a carrot-haired flyer – John Gallagher.

For four years and in 50 matches they never tasted the bitter pill of defeat. The closest they came was a 19-19 draw against Australia in 1988. So high was the standard they set themselves, they treated that as a loss.

For four years and in 50 matches they never tasted the bitter pill of defeat. The closest they came was a 19-19 draw against Australia in 1988. So high was the standard they set themselves, they treated that as a loss.

"I was devastated," said Sean Fitzpatrick later. "I probably had only a couple of bad throw-ins that day but I thought my performance was utterly hopeless. The other guys felt the same way. We considered ourselves failures."

But all unbeaten runs must come to an end sometime. The Aussies and others had been catching up. They came to tour in 1990 and by that time there had been changes in the great team. John Drake had retired. John Gallagher had been lured to league with big money. Michael Jones was out recovering from injury. Joe Stanley had been dropped. So had Buck Shelford in controversial circumstances when he had been captain and the team still unbeaten.

The changes seemed to work though. The Aussies were hammered 21-6 in the first test. But they crept closer to 27-17 in the second test. With the series won and the Bledisloe Cup retained, the hunger went out of the All Blacks, while the Australian side that would win the World Cup a year later, were desperate for a win.

On the day of the third test the weather was ominous with overcast skies. The game was as about as exciting as a funeral with only one try scored – to Aussie hooker Phil Kearns. The Wallabies won 21-9, finally breaking the All Black spell.

Though it was painful for all New Zealanders to bear, it did put into perspective the great string of victories that had preceded it.

77

JONES THE TRY

1. 22 May 1987 – Eden Park, Auckland. 2. 3 October 1991 – Twickenham, England

The Iceman sets a record that no one will ever eclipse

Great players stand out by their deeds and there are some who say Michael Jones is the greatest player of them all. He played in two World Cups and could have possibly played in three except for his no-play-on-a-Sunday stance. In the two he did play in, he created a unique record.

22 May 1987 – Eden Park, Auckland

It was the opening game of the first Rugby World Cup between New Zealand and Italy. It was also the first international of the year for the All Blacks. Nobody was certain how well this team would play.

The team was young but there was only one new cap, an athletic open-side flanker from Auckland called Michael Jones. The Italians attacked first, but the All Blacks withstood the assault. They forced several scrums and finally the Italian captain Innocenti tackled All Black number 8 Buck Shelford when he did not have the ball. Penalty try. The All Blacks had the points, but a real try has not yet been scored in the World Cup.

Then the All Blacks won a scrum from the Italian put-in. Buck Shelford broke away around the blind-side. He passed to Alan Whetton who pushed the ball on to Grant Fox who passed it to Jones. There were players outside him but Jones could sense the closeness of the tryline and decided to go for himself. The crowd roared as he crashed over to score the first real try of the first World Cup.

The crowd roared as he crashed over to score the first real try of the first World Cup.

3 October 1991 – Twickenham, England

Four years later it was the opening game of the 1991 World Cup. The champions New Zealand faced the host nation England. Michael Jones was no longer the new kid on the block, though he sometimes felt like it. Two years before his career seemed to be finished when he injured his knee against Argentina. He spent almost a year recuperating. Drawing on great courage he fought his way back to fitness and regained his position in the All Blacks.

England's big grinding pack tried to beat the All Blacks in the forwards. The match was not pretty, as Grant Fox and the English fullback Jonathon Webb participated in a kicking duel. At halftime Webb was ahead by one as England led 12-9. The game was crying out for a try.

Eleven minutes into the second half, the ball came back to Graeme Bachop from a scrum just outside the English 22. He tossed a short pass to Craig Innes then doubled around him to take the ball back again before throwing out a long pass to John Kirwan. J.K. sprinted past the English wing Oti, but with the try-line beckoning was halted by the covering tackle of Webb.

Big strong J.K. was able to stand in the tackle and turned to glance at his shadow Michael Jones. Jones took the pass and crossed the line unopposed.

For the second time in a row Michael Jones scored the first individual try of a World Cup tournament.

78

CENTENARY TEST

22 April 1992. Athletic Park, Wellington

The NZRFU has 100 candles on its cake

The New Zealand Rugby Football Union was formed in 1892 in slightly inauspicious circumstances. The "founder" was 27-year-old Edward Hobden, who toured the country during 1891 trying to rustle up enough support among the local unions for a national body. When the inaugural meeting of the NZRFU was eventually held at the Club Hotel Wellington in 1892 the oldest union, Canterbury, was absent. Otago and Southland also refused to affiliate at that time.

One hundred years later there were 27 unions throughout New Zealand. As part of the centenary celebrations a World XV was brought to New Zealand to play three matches against the All Blacks. The invitation team included a diverse array of some of the best players in the world included Naas Botha (South Africa), Nick Farr-Jones (Australia), Gavin Hastings (Scotland) and Peter Fatialofa (Western Samoa).

The centenary could not have come at a worse time for New Zealand rugby. The previous year the All Blacks had lost the World Cup amidst talk they were more interested in the commercial spinoffs than wearing the silver fern with pride. There was a new coach in Laurie Mains who was forced to discard some of the former stars and bring in new, untried players. His captain was due to be Mike Brewer, but Brewer was injured before the first game and Sean Fitzpatrick took over.

The whole weight of the history of New Zealand rugby sat on the shoulders of this young All Black team.

As captain, Fitzpatrick got a room of his own. When he mentioned on the radio how lonely this was, someone sent him a great big teddy bear. The new look All Black side played like teddy bears in the first test, losing 28-14, the second highest score posted against New Zealand in a test. Even worse, a host of former All Blacks sat in the stands watching, including Jock Richardson who had captained the 1924 Invincibles in their tests against Ireland, Wales and England and at the time was the oldest living All Black.

There were six changes for the midweek test against the World XV. Established All Blacks like John Kirwan, Michael Jones and Grant Fox were out. The night before the game the players attended a function where the Prime Minister Jim Bolger thanked the All Blacks for being so generous in allowing the visitors to see so much of the ball. He told them it was such a shame that the All Blacks who used to be

World Champions could not even beat a scratch collection of players thrown together just for a few games.

The whole weight of the history of New Zealand rugby sat on the shoulders of this young All Black team. Lose this match and they might as well follow the World XV players back overseas (except for Peter Fatialofa who was a South Auckland piano mover). They would not be able to show their faces in New Zealand again.

Eight minutes into the match the All Blacks had an unexpected "bonus" when the French lock Roumat was sent off for kicking Sean Fitzpatrick. The All Blacks were already ahead and began to open up, with tries to Loe, Cooper and Clark. As the World XV desperately tried to fight back Vaiga Tuigamala intercepted a long pass from Guscott to Claassens to score a thrilling 70-metre try.

In all, the World XV was nailed for ten tries in a final score at 54-26. New Zealand pride had been restored. The All Blacks went on to win the final game in slightly less spectacular fashion 26-15 to clinch the series 2-1. All the former All Blacks in the crowd had smiling faces again.

79

TRIUMPHANT RINGING OF COWBELLS

4 October 1992. Rugby Park, Hamilton

Waikato rampant in first ever NPC play-offs

Until 1976 rugby matches in New Zealand were technically "friendlies" though this did not describe the way most New Zealand forward packs traditionally went about their business.

With the birth of the NPC each match took on a new significance with two points at stake. But even so, the season had a way out fizzling out towards the end, with the championship winner often already decided well before the final matches were played.

In 1992 it was decided to adopt "play-offs." The concept was immediately successful when fourth-placed Waikato went to Auckland and shocked the championship-winning team at Eden Park. Only 10,000 people turned up for the match. Perhaps Aucklanders were saving their money for the final. They shouldn't have

...a close match could have been expected, but Waikato played sublime and flawless rugby.

bothered as Waikato pulled off a shock 27-21 win.

The next day at Carisbrook an even more exciting semi-final took place between Otago and North Harbour. When the whistle blew for fulltime both teams were locked together at 16-16. According to the rules that meant both sides had to pull themselves up for 20 minutes of extra time.

North Harbour looked as if they would secure the match when Ant Strachan scored a try. It got worse for Otago when the try was converted, but Greg Cooper clawed the score back with a penalty. Then Otago launched a series of thrilling raids which culminated in a try to Greg Cooper in the corner. He didn't need to convert it, Otago were through to the final. But he did anyway to make the final score 26-23.

In the final, Cooper's reward was the infamous eye-gouging incident by Richard Loe which detracted from a superlative performance by Waikato who had effectively come into the play-offs as the underdogs. Especially after the semi-finals a close match could have been expected, but Waikato played sublime and flawless rugby. Their forwards pushed the Otago pack onto the back foot, allowing the backs to clean up 23-0 at halftime. Coach Kevin Greene and captain John Mitchell warned them in the interval that it was not enough.

In the second half Waikato roared away to a 40-0 lead before Otago's Stu Forster scored a consolation try to make the score 40-5. They had become the first NPC play-off champions by completely blowing away the southerners. The bell ringing crowd of 30,000 was ecstatic.

80

BACK TO AFRICA

15 August 1992. Ellis Park, Johannesburg

Zinny too clever for Boks

In 1992 there was a whisper in the air that stunned most rugby followers in New Zealand. With the proposed changes in the political system in South Africa the Springboks were going to be allowed back into world rugby. As soon as it was confirmed, South Africa extended an invitation to New Zealand to tour. With equal alacrity the New Zealand Union accepted.

The last time the two old foes had officially faced each other was in the unforgettable flour bomb test of 1981. Eleven years and two World Cups had passed. No All Blacks had survived in the test line-up from that last meeting, but two South African greats had in Naas Botha and centre Danie Gerber. Despite, and perhaps because of, their many years of isolation, the South Africans considered themselves the "World Champions". They claimed the first two World Cups had been won by others only because they hadn't been there.

...the political situation was still tense. Everywhere they went, the All Blacks were protected by armed guards.

It was midnight when the All Blacks flew into Johannesburg airport to be greeted by several thousand people. Though the reception was friendly, the political situation was still tense. Everywhere they went, the All Blacks were protected by armed guards.

New Zealand easily won all four games against provincial teams and then travelled to the altitude of Ellis Park, Johannesburg for the test. Nobody knew quite what to expect. It had been 11 long years.

After 14 minutes Grant Fox opened the scoring with a penalty. In the second half the All Blacks started to pull away. That was followed by a try to Vaiga Tuigamala when he picked up a Bok kick and charged back to set up the ruck. The ball was thrown out to Frank Bunce who sliced through the South African defence and sent the perfect pass to John Kirwan who scored under the posts. Soon afterwards John Timu scored in the corner, and Fox was able to convert with a superb kick.

With five minutes of the game left the All Blacks were 17 points ahead. Then the altitude and All Black's long season started to kick in. The Springboks made a spirited recovery and scored two quick tries. But it was too late as they ran out of time in an exciting finish.

The deciding factor in the match had already occurred in the 33rd minute of the first half when South Africa were penalised for being offside at a ruck near their own line. As Grant Fox walked towards the ball, the South Africans turned and trouped back to behind their line to await the kick. They had been out of test match rugby too long. In contrast Zinzan Brooke was battle-hardened, his natural instincts honed to the test match arena.

In an instant he picked up the ball, tapped it to himself, charged at the startled South Africans and threw himself over the line for a try which Fox converted. It was the crucial score in a final victory of 27-24. After all their years of seclusion, the South Africans had experienced nothing to prepare them for the crafty Zinzan Brooke.

81

A STAR IS BORN

3 April 1993. Levin Domain, Levin

A try that dreams are made of

Nearly every New Zealand boy dreams of being an All Black. But they are not just misty thoughts, they are very real, enacted in the delirium of youth on many windswept rugby paddocks.

As the barefoot youngster sidesteps the daisies, he dreams of being plucked from school to play against a visiting team of renown with such rare brilliance that all sit up and take notice. He has the ball in open play and faced by the man who will soon be the All Black fullback he…and there the dream starts to fade into mist, because surely *that* would be too unreal.

But it really happened to Carlos Spencer. In 1992 he made his debut for Horowhenua as a 16-year-old schoolboy. At first he felt "weird" being a youngster amongst the men.

The following season, 1993, was Horowhenua's centenary and in acknowledgement of this the Ranfurly Shield holders Auckland agreed to accept a challenge from the third division union on their early season tour. To most of New Zealand it seemed that Auckland had held the Shield "forever" though it had been only eight years. They had begun to think of it as permanent Auckland property.

The match against Horowhenua on the 3rd of April was the earliest time in the year the Shield had been challenged for and it seemed half of Levin were there to watch the game and enjoy the sunshine.

It was too much to believe that Horowhenua could win the Ranfurly Shield but they did have their day of glory…

Facing Horowhenua was what was to become the All Black front row of Olo Brown, Sean Fitzpatrick and Craig Dowd and in the backs was "the big black bus" Inga Tuigamala. Directly opposing Spencer at first five-eighth was the legendary Grant Fox. They were a study in contrasts. Fox was methodical and efficient while Spencer was all talent and instinct. Fox wore his hair short and neat while at the time Spencer sported dreadlocks.

Spencer was excited as the game kicked off. "I was thrilled to be out there on the same field as those guys. Just to be out there playing against them was something else."

As expected, Auckland used their steamroller pack to take control of the game and release the backs who scored 11 tries, four of them going to Lee Stensness.

It was too much to believe that Horowhenua could win the Ranfurly Shield but they did have their day of glory with two stunning tries. Spencer made a break for one of them and the ball was passed on to winger Nopera Stewart who sprinted 60 metres to score in the corner.

For the other try Horowhenua took a quick tap from 45 metres out. A reverse pass sent centre Craig Yorston bursting through and he passed to Spencer who was backing up. He went through a gap and then was faced by Shane Howarth who was shortly to become the All Black fullback. As Howarth set to tackle, Spencer threw a dummy. Howarth bought it and Spencer stepped past him to sprint to the to the line and score the try that dreams are made of.

82

INTO THE LIONS' DEN

3 July 1993. Eden Park, Auckland

The unbearable pressure of test match rugby

The word "test" was derived from the ancient practice of heating up a metal to a great heat to test its purity. It is no coincidence that the word is used to define an international rugby match. Here is the cauldron in which players who have performed well in the NPC (and from 1996 the Super 12) are tested to the very edge of their limits. So intense is the searing cauldron of test match rugby that it will make them or break them.

The heat never got any hotter than during the British Lions tour of 1993. The first test of the three-match series was close throughout with the Lions playing a tight, efficient game. New Zealand scored one try, the rest of the points in the match coming from kicks. With three minutes to go and the Lions leading they were penalised at a maul. "You've kicked this thousands of times before in training. It's just another kick," Grant Fox told himself as he stepped forward to take it from 42 metres out. A nervous twitch of the fingers, a couple of steps and he thumped the ball over to give the All Blacks a lucky 20-18 victory.

In the second test the All Blacks played miserably and were hammered 20-7. Sean Fitzpatrick considered it the worst game he had ever played. He gave away two penalties, both of which were kicked by the Lions. When he knocked the ball

Fitzpatrick was there to lead the All Blacks out onto Eden Park, his face a stone mask of concentration.

NEW ZEALAND HERALD

on, the Lions scored from it.

Over the next few days the New Zealand media went mad, frothing at the mouth to rubbish this All Black side that was being outplayed by a relatively mediocre Lions unit.

The tension going into the final test was almost unbearable. The younger players could hardly stand it. The older players knew that if they lost it was not only the series but also the end of them as All Blacks. The pressure was so bad Fitzpatrick even wondered if he would be picked for the test. At night he tossed and turned in his sleep, again and again, hearing the frustrated words Grant Fox had uttered when Fitzpatrick had knocked on in the second test, "That's the series."

But Fitzpatrick was there to lead the All Blacks out onto Eden Park, his face a stone mask of concentration. There were three changes to the team and new tactics – to open up the game and tire out the Lions' big forwards.

But the tension caused the All Blacks to misfire at the start. The Lions took the lead with a penalty and then were 10 points ahead with a freakish try. The Lions' wing Underwood cut back into the midfield where he was tackled by Kirwan. He attempted to pass towards the other wing Evans, but the ball struck Frank Bunce on the knee and careered towards the All Black line where it was gathered up by Scott Gibbs who went over for the try.

So far the Lions had managed to beat the All Blacks with committed defence. As the All Blacks went on the attack in the Lions 22, a flat line of defenders came up to meet them. New cap Lee Stensness sensed the situation perfectly and chip kicked over the head of the Lions' defence for Frank Bunce to run on and score. Up to that point it was only the second try the All Blacks had scored against the Lions in the entire series. It was the breakthrough and the Lions crumbled after that, letting in two more to Fitzpatrick and Preston. The All Blacks won 30-13 and the pressure was off.

You can taste the tension as the Lions and the All Blacks battle for possession in the deciding test at Eden Park 1993.

83

MOOLOO MAGIC

18 September 1993. Eden Park, Auckland

Waikato end Auckland's record-breaking Shield tenure

Although Auckland were willing to take the Ranfurly Shield out on tour to let the smaller unions such as Horowhenua and Buller have a wink at it, they made sure it was safely kept at Eden Park under lock and chain when the more powerful unions came to challenge. Though there were some in the Queen city who said

NEW ZEALAND HERALD

The strain of defeat is already etched on Grant Fox's face, the concentration of victory on Ian Foster's, the day Waikato won the shield.

The crowd was awash with red, yellow and black and the air was filled with an orchestra of cow bells, ringing so hard that the Auckland players could hardly hear themselves think.

it was hardly worth taking it out of the trophy cabinet and displaying it on the field. Out there it stood the chance of warping in the sunlight or being attacked by rust. Best to keep it safe where it belonged.

Around the country some of the other provinces were starting to lose interest in the Ranfurly Shield. "What was the point of going all the way up to Auckland to get a 40-point pasting?" they were asking. But down in Waikato country they were starting to hanker for it. The men from Mooloo were getting a little bit tired of being the poorer neighbours of New Zealand's largest metropolis and started hatching a plan of how they could snatch the Shield.

First of all the whole Waikato front row of Craig Stevenson, Warren Gatland and Graham Purvis cut their hair in a new aggressive bristle headed "Waikato" haircut. The next part of the plan was simply to take over Auckland. The Auckland public had grown blasé about holding the Shield and often preferred to fix up the garden or wash the car on a Saturday afternoon.

Convoys of Waikato fans made the trip up the motorway, converging on Eden Park and virtually turning it into a home game for the Mooloos. The crowd was awash with red, yellow and black and the air was filled with an orchestra of cow bells, ringing so hard that the Auckland players could hardly hear themselves think.

On the paddock Waikato went straight onto the offensive with a reverse kick off. From there they never let up, tackling like tigers, aggressive and committed in the set pieces and kicking cleverly. They were ahead after five minutes with a penalty to Foster.

Fox got one back, but when Foster scored a try that took the challengers out to 8-6 at half time. Surely it would not be enough against a team that had held the Shield for eight years. But Waikato kept on the attack, pushing further ahead with three penalties to Matthew Cooper to win 17-6.

At the end of the match long-serving Waikato captain John Mitchell got to hold the Ranfurly Shield aloft, the first person outside Auckland to do so in a long while.

The Auckland players sportingly congratulated the victors. "It's a bit of a weight off the shoulders," said John Kirwan. "But not a weight that you want to get rid of."

That evening when the long convoy of vehicles headed back towards Hamilton the "log o' wood" went with them.

84

GOLDIE'S TEST

20 November 1993. Murrayfield, Scotland

Jeff Wilson makes a stunning test debut

True or false – Jeff Wilson first played for New Zealand in an international match on 20 November 1993? True, the Scottish rugby historians would reply ruefully, wishing it wasn't true. But they'd be wrong. Before he made his debut for the All Blacks Wilson had already represented New Zealand at cricket.

In 1993 after a successful season playing rugby for Otago Wilson was looking forward to the cricket season again. He knew he couldn't make the All Blacks. On one wing was the pocket battleship Vaiga Tuigamala and on the other the legendary John Kirwan. Kirwan was also looking forward to a change of sport when some friends gave him a new surfboard. He got more use out of it than he expected when he was sensationally axed from the All Black tour to England and Scotland at the end of the season.

Meanwhile 19-year-old Wilson was also shocked to be told that he was going on the tour and had to put his cricketing plans on hold. Originally the All Black coaching panel had selected Wilson to play in the minor games and be a "learner".

Wilson was a fast learner. While still at school he had scored 66 points in one rugby match. He was only 15 when he had made his debut for Southland at cricket and one leading basketball coach considered that if Wilson had carried on in that sport he would also have become a Tall Black.

...after a successful season playing rugby for Otago Wilson was looking forward to the cricket season again. He knew he couldn't make the All Blacks.

In 1993 Laurie Mains judged his progress good enough, just a few weeks after his 20th birthday, to throw him straight into the fray against Scotland at Murrayfield. The stand was being rebuilt at the time and the British press found themselves sitting closer than normal to the action. So close that they felt the shudder as the New Zealand forwards drove into the rucks to set up an effective platform for their backs.

Wilson was soon into the action, almost scoring a try in the first 15 minutes when he was stopped just short of the Scottish line. The ball was moved wide from the breakdown and Ellis scored instead.

Then Wilson was put into space by an overhead pass from Frank Bunce. He swerved around the Scottish centre Jardine to score his first test match try. His second came from a kick ahead by John Timu as Wilson chased, gathered the ball and touched down. By now Scotland was well beaten. As the last moments of the match ticked away, the All Blacks won the ball from a ruck and Forster passed the ball out to Wilson who was hungry for more. He sprinted away from the cover defence to score in the corner for a memorable hat trick.

Mathew Cooper had kicked six goals for the All Blacks that day but was forced to leave the field injured. That left Wilson to attempt to convert his own try. It was a difficult kick right from the sideline.

Golden Boy couldn't do anything wrong. The ball flew straight and true between the posts to rack up New Zealand's half century (51-15).

85

COASTERS TRIUMPHANT

24 September 1994. Rugby Park, Greymouth

West Coast finally wins a match

Before a tough test one time, All Black Bernie Fraser was heard to say, "If this game is about anything, it's about winning." But at the other end of the spectrum it's mostly about losing.

West Coast provincial rugby was in the doldrums. They were one of only four unions who had never won an NPC title. Worse than that, they couldn't even win a match.

One of the perennial backblocks of New Zealand rugby has been the West Coast. Actually what is geographically known as the "West Coast" of the South Island contains two quite separate unions West Coast and Buller. Both are small and linked by one winding coastal road. Though they have often flirted with amalgamation, neither has yet been willing enough to give up their independence.

West Coast is also one of the oldest unions. It was formed in 1890 but did not affiliate with the NZRFU until 1893. The West Coast Union has always had the added problem of having to compete with rugby league, which is very popular in the area. Though in times gone by, when rugby power was less concentrated in the metropolitan areas in New Zealand, West Coast has provided seven All Blacks. The most famous of them was the professional sprinter Jack Steel who played for the 1924 Invincibles and who scored that classic try against the 1921 Springboks.

In recent times though, the West Coast has struggled to put out a competitive team as the heavy industry in the area has closed down and the young men have left in search of work. Since the 1980s the population in the region has been dropping and this has shown up in the results on the rugby field.

By the beginning of the 1990s West Coast provincial rugby was in the doldrums. They were one of only four unions (along with Buller, East Coast and North Otago) who had never won an NPC title. Worse than that, they couldn't even win a match.

The 1994 season didn't start off any better. In the first match away to Thames Valley they were thumped 74-5. They came closer against Marlborough losing by just one point. The margin of defeat against Mid-Canterbury was a mere two points.

In September when North Otago came to visit, West Coast had lost 33 consecutive NPC matches. It didn't look too promising for West Coast because North Otago had recently slaughtered East Coast 48-29, scoring seven tries. For five years West Coast had sat at the bottom of the New Zealand provincial rugby ladder, propping up the third division in each of those seasons. They were desperate for a win. Maybe the weather was on their side, because the game was a typical West Coast squall that turned the field into a quagmire.

The team played their hearts out in the mud and the rain and that day as the West Coasters achieved their own World Cup – they finally won a game again. The 24-7 victory lifted them off the bottom of the table, finishing two points above North Otago whose turn it was to prop up the other 26 teams in the NPC.

86

WILSON'S DIVE

17 August 1994. Sydney Football stadium, Sydney

The try that almost was

The next split second of action was one of the most memorable to occur on a rugby field.

For the first three years after the Laurie Mains took over the All Blacks in 1992 their form was patchy. They struggled to beat the scratch World XV, almost lost to a poor Ireland side, stumbled against the unimaginative Lions and then got whitewashed by the over-imaginative French.

With the World Cup coming up the following year, it was hoped that in the last match of 1994 against Australia in Sydney for the Bledisloe Cup, the All Blacks would deliver an outstanding performance. But they stuttered from the start.

Fullback Shane Howarth had just completed the haka and was still getting his focus in on this night match when he saw the ball spiralling towards him. The Aussies had kicked off, the All Blacks had failed to claim it and now he was expected to catch this bomb. As Howarth stood there waiting for it Jason Little flew above him to grab the ball and score. Just 20 seconds had elapsed. Not a good start at all.

New Zealand did not mount their first attack against the Australian line until 11 minutes into the game. It was all Australia as they racked up a half-time lead of 17-6. "You've played harder than this in training," Laurie Mains told his players at the interval. They were not even breathing hard. In the second half the All Blacks didn't just pick up the pace, they raised it to a blistering level the Aussies just couldn't live with. Wave after wave of black attacks stormed into Australian territory leaving their opponents gasping for time and breath.

Seven minutes into the second half Michael Jones dived over the Australian tryline into the left corner, but referee Ed Morrison judged Graeme Bachop's pass to be forward. Three minutes later it was desperate Australian defence that stopped Howarth scoring in the corner. The All Blacks lifted the pace even further and again were only just repulsed before the desperate Australians conceded a penalty which Howarth kicked.

But no one could live with this pressure. Mathew Pini the Wallaby fullback was caught on his 22. Frank Bunce led the attack and passed the ball to Howarth who danced inside two tackles to score.

Howarth converted his own try. A penalty by Knox took the scores out to 20-16. The All Blacks swept onto the attack again. The Australian defence seemed to have the situation covered when the ball came out to Jeff Wilson on the right wing. He moved in then out in a sweeping movement which created space where there had been none. The Aussie defence melted before him, suddenly there was only grass between him and the tryline. He began to dive as he crossed it.

The next split second of action was one of the most memorable to occur on a rugby field. It was the meeting of two players with two techniques that exactly contrasted each other. Wilson's practice was to dive high over the tryline, the showman in him celebrating the try with a final spectacular flourish. But tackling Wilson as he crossed was the "black opal" George Gregan, one of the greatest cover tacklers in the world, who had perfected one particular type of tackle where he attacked not the player, but the ball.

As Wilson dived for the certain try that would win the game, Gregan's hand came through and knocked the ball from his grasp. They fell to the ground together, shock and horror on Wilson's face as the ball flew away over the dead ball line.

Afterwards Wilson was in tears in the changing room. The match was lost, but out of the ashes of defeat the All Blacks developed a style that would carry them forward into the following year's World Cup – play the game at pace.

87

THE STOUSH OF ONEWA DOMAIN

15 October 1994. Onewa Domain, Takapuna

The first rule of rugby

Rugby is war with more rules and fewer casualties. New Zealanders have always played a hard form of rugby that has occasionally prompted visiting tourists to return home complaining that rugby is supposed to be "fun." But New Zealanders are having fun, knocking chunks out of the opposition and anyway, the first rule of the rugby paddock is, "no mates and no prisoners taken."

In 1994 Auckland were performing below their usual high standards. They were forced to win their semi-final match away against Otago to make the final. There they had to play away again. North Harbour had finished higher on the championship table and therefore had home advantage.

Before the new North Shore stadium was opened at Albany in 1997, North Harbour played at Onewa Domain, a small ground that usually gave an intimate atmosphere with the crowd close to the touchlines. This match should have been an "intimate" affair as both North Harbour and Auckland were loaded down with All Blacks who had just finished a season playing together.

Auckland and North Harbour are also geographic neighbours separated only by a yawning harbour and a long bridge. The week before the game signs started going up on the bridge with Harbour supporters *politely* informing Aucklanders what would happen to them after they had crossed it. Some radio stations also got some needling banter going during the week.

On the day of the game 14,000 people squeezed into tiny Onewa Domain expecting to see an exhilarating game of rugby. Instead they saw the dark undercurrent of the sport that is often hidden from the cameras. The close rivalry of the two Unions, the claustrophobic atmosphere of the ground and perhaps the fact that Auckland had been struggling while Harbour had been playing well, all added to a mix that exploded into civil war on the field.

...14,000 people squeezed into tiny Onewa Domain expecting to see an exhilarating game of rugby. Instead they saw the dark undercurrent of the sport.

NEW ZEALAND HERALD

Zinzan Brooke is happy despite being felled by Eric Rush and also seeing brother Robin being sent off. Auckland has literally won a hard fought NPC final.

Both sides had players sinbinned as fists, boots and elbows flew. Robin Brooke was sent off. Eric Rush of Harbour took exception to something his fellow All Black Zinzan Brooke did and poleaxed him. Rush also got his marching orders. It was definitely "no mates and no prisoners taken."

Amidst it all a little bit of rugby did get played. North Harbour was ten points down at half time, but closed back to 10-9 with three penalty goals to Warren Burton. But then Burton made the error that won Auckland the game, fumbling a pass from Ant Strachan near his own line. Sean Fitzpatrick snaffled up the ball and scored the try. Auckland scored again through replacement Chandler and despite a try to Walter Little to bring the score back to 22-16, it was too late. Auckland won their second consecutive national championship.

But not many people left the ground remembering the rugby. There had been too many punches thrown for that. The following week Robin Brooke and Eric Rush appeared before the judiciary. Brooke was banned for one month. Rush got a lesser sentence of one week, apparently because he would shortly be needed to captain the New Zealand Sevens team.

88

THE RHINOCEROS IN BALLET SHOES

25-26 March 1995. Hong Kong Stadium

Jonah thrills the crowds at Hong Kong Sevens

By the time of the 20th tournament in 1995 the Hong Kong Sevens had grown to become the biggest event in the world of sevens rugby, a weekend extravaganza of running rugby, tries, more tries, old friends and more old friends.

From the first 12 teams drawn originally from Oceania the competition had expanded to include teams from rugby playing countries all around the world. With the increased numbers, after the first seven years the tournament had moved on from the cramped but family atmosphere of the Happy Valley Ground to the sprawling expanse of the Hong Kong Government Stadium.

New Zealand teams had been coming to the tournament since the very first one had been won by the Cantabrians club of Christchurch in 1976. In the following two years New Zealand sides were runners up. Marlborough lost to Fiji in the final in 1977 and then Manawatu lost to the Fijians again in 1978. For many years New Zealand persisted in sending the national provincial sevens champions instead of a national team. In 1982 there was no invitation issued to a New Zealand team and the following year the NZRFU took the hint and organised a national sevens team – who were promptly invited.

It was 1986 before New Zealand won the sevens, a feat repeated in 1987 and 1989. In 1990 and 1991 they were runners up to the powerful Fijians. New Zealand won again in 1994. But it was the Fijians who were considered the kings of sevens rugby. It was their national sport, sevens are far more popular in Fiji than 15 a side, and didn't they have the all-time master of sevens in Waisale Serevi?

In 1995 the canny Gordon Tietjens was the coach and on the field the team was led by the old sevens fox Eric Rush. Other members of the squad were Andrew Blowers, Brad Fleming, Dallas Seymour, Joe Tauwi, Joeli Vidiri, Peter Woods and a youngster by the name of Jonah Lomu.

Though new to the Hong Kong Sevens Jonah had actually been playing international sevens since he was a fifth former at school.

Though new to the Hong Kong Sevens Jonah had actually been playing international sevens since he was a fifth former at school. After he had outpaced Eric Rush in a sevens tournament in 1991, Rush invited him to join his Mongrels team for a tournament in Singapore.

But when he arrived at Hong Kong in 1995 Jonah was overawed by the size of the tournament and the crowd. It wasn't long before opposing teams were overawed by both Jonah's size and his speed. He seemed to be even faster from a standing start. For one of his tries during the tournament he actually slowed down to a stop and when his opponent did the same, Lomu took off like a greyhound out of the trap, running around him and beating him to the try line.

In the final the New Zealand team beat their old rivals Fiji 35-17. Jonah was made player of the tournament. Certainly the image most people took away with them was of the thundering runs of a 1.95m rhinoceros in ballet shoes.

89

JONAH'S MATCH

18 June 1995. Newlands, Cape Town.

Big, brown, brooding Gulliver careers through the English

Jonah Lomu almost didn't make it to the 1995 World Cup. As a teenager he turned down a $100,000 offer to join Canterbury Bankstown League Club. Then he was dropped from the All Blacks in 1994 after being the youngest ever player to pull the black jersey over his chest (at 19 years and 45 days). Inexperienced as a winger, he was out of his depth against the crafty French. He fought back to make the gruelling summer training camps, then almost went to league again when a false news item appeared on TV3 that he was out of contention with the All Blacks.

At the warm-up game against Canada prior to the World Cup Marc Ellis had played on the left wing, but Lomu was in for the tour opener against Ireland. None of the opposition knew what to make of this thundering giant. They backed off him, gave him room and he ran around them. The All Blacks were easily the best team of the tournament as they came to the semi-final against England.

The experienced players in the party recalled the last match the two teams had played at Twickenham in 1993. The English had played with three no 8s in the back

NEW ZEALAND HERALD

The England players reverently look down at the ground as the God-like Jonah Lomu runs in his fourth try in the 1995 World Cup semi-final.

row to negate the All Black's driving tactics. They would not be an easy team to beat.

All week the All Blacks psyched themselves up for the match, then they got some help from an unexpected source – the English wing Tony Underwood. Perhaps trying to talk up his own confidence, he told the media that Jonah had never been tested defensively.

The newspaper reports of those comments were stuck to Jonah's door by the All Blacks – just in case he had missed them. Someone should have warned Tony Underwood to keep his mouth shut. The last thing you want to do is upset a 1.95m tall, 116kg, taro-eating, Tongan All Black.

The night before the match Jonah didn't sleep at all and says he played the next day on, "sheer adrenalin." He was supercharged as he ran out onto the pitch,

Someone should have warned Tony Underwood to keep his mouth shut. The last thing you want to do is upset a 1.95m tall, 116kg, taro eating, Tongan All Black.

hungry for the ball and hungry for the try line. He and the tens of thousands of spectators at Newlands did not have long to wait.

When New Zealand got hold of the ball after just two minutes, Graeme Bachop flung out a wild pass that flew over Jonah's head. Jonah turned to catch the bouncing ball, then shrugged off the tackle of Tony Underwood and sprinted around the despairing dive of Will Carling. Between Jonah and the line was English fullback Mike Catt, but so intent was Jonah on scoring that he did not even see him and trampled right over the top of him.

He went on to score three more tries in the most dominating individual display of rugby anyone had ever witnessed. The English were shell-shocked; the rugby world was awestruck. It was a game that no one who saw it would ever forget.

The next day the newspaper headlines were full of giant Jonah. Peter Jackson summed it up best in the London *Daily Mail* when he wrote, "The Polynesian warrior wing careered around Newlands like a brown, brooding Gulliver in the land of Lilliput..."

90

TEARS AND DROP KICKS

24 June 1995. Ellis Park, Johannesburg

The effort was valiant, but the All Blacks opposed the re-birth of a nation

For the first time a Rugby World Cup final had to go into overtime.

New Zealand had won the very first World Cup in 1987. Afterwards the celebrations had been very dignified. They had expected to win it. In 1991 they were expected to win it again, but were ambushed by a very good Australian team who were aided by the wizardry of David Campese. Then they had stuttered through the next three years as Laurie Mains blooded players and the All Blacks tried to develop a style.

They had found it in the middle of a crisis, while losing the test against Australia at the Sydney Football Stadium in 1994. Play the game at pace. No one else could live with that. They had done that at the World Cup and with Jonah Lomu, a ballistic missile on the wing, they had blown all the opposition away.

South Africa had missed the first two World Cups because of apartheid. They

NEW ZEALAND HERALD

The exultation of the South Africans is heightened by the grim faces of the All Blacks at the final at Ellis Park.

suffered problems early in the tournament when two of their players were banned as a result of an all-in brawl against Canada. They had scraped through in the semi-final against France, a last minute attempt at a try by the French falling a mere metre short.

On performance the All Blacks were the better team, but South Africa were playing at home and they had a whole nation behind them. Their spirits were lifted even higher when the new President Nelson Mandela came onto the field wearing the same number 6 shirt as the South African captain Francois Pienaar.

With great spirit and superb defence the South Africans did what all the other teams had failed to do so far – they closed down Jonah and the rest of the New Zealand running game.

The match became a tight, tense struggle with the only points scored from the boots of Joel Stransky and Andrew Mehrtens. The lead changed hands, but neither team could break away from the other. With just three minutes of playing time left and both teams seeming to have fought each other to a standstill, the All Blacks had a scrum thirty metres out. The ball was sent back to Andrew Mehrtens who took a drop kick at goal. The ball flew high – but wide.

For the first time a Rugby World Cup final had to go into overtime. As players on both sides pushed their tired bodies to the limit of fitness, New Zealand scored first with a penalty which was answered by one from Joel Stransky. As the second half of extra time kicked off the teams were still level.

With seven minutes to go, South Africa put pressure on the All Blacks and the ball came back to Joel Stransky who readied himself and kicked. Behind the goal the whole crowd seemed to be a sea of triumphant hands raised into the air as the ball flew between the posts.

When the final whistle blew the tough lock Robin Brooke shed tears of defeat. Despite being the outstanding team of the tournament, the All Blacks had failed to win the World Cup.

91

PLAY FOR PAY

1 March 1996, Palmerston North Showgrounds

The very first professional Rugby Match

Rugby was the last great sport to grasp the coin of professionalism. From the very start an amateur ethos had pervaded the game. There was shock and horror in 1893 when a group of cloth-capped Englishmen claimed players should be paid for *lost time.* So they left and formed their own Northern Union which was later renamed Rugby League.

Rugby has been traditionally administrated by plodding dinosaurs who would have dawdled along for another century of amateurism if it had not been for Australian (now a US citizen) media magnate Rupert Murdoch's apparent desire to hold the TV rights to virtually every sporting contest going. A money war with Kerry Packer to obtain the rights to Australian League threatened to steal away many of rugby's best players so the dinosaurs blinked into the sunlight and rugby went professional.

The South African, Australian and New Zealand Unions (SANZAR) signed a deal with Murdoch to provide an early season tournament which included the 12 best provincial teams – the Super 12. This historic first game of the new professional rugby era took place at Palmerston North Showgrounds on 1 March 1996 between the Wellington Hurricanes and the Auckland Blues. A crowd of 13,000 people turned up to watch the match which began with a fireworks display before the kick-off.

There were more fireworks straight after the kick-off as the Hurricanes tried to blow away the Blues. After just two minutes the Hurricanes forced a penalty near the Auckland line. They took a quick tap and Alama Ieremia darted through a half gap to score the try – the first of the professional era.

The Blues gave away more penalties as they came under more pressure from the force of the Hurricanes, resulting in three successful penalty kicks by Jamie Cameron. After 30 minutes the Blues still had not got any points on the board and the Hurricanes' lead was out to 16.

...the dinosaurs blinked into the sunlight and rugby went professional.

Finally the Blues got some continuity going and after bringing the ball up field from the half-way line scored with a try to Andrew Blowers. When Eroni Clarke scored after a quick throw-in from Ofisa Tonu'u that brought the scores back to

16-all. The two sides thrust back and forth like fencers, one scoring, then the other, each vying to be the side to win the first professional game of rugby.

With three minutes to go the Hurricanes still led but the Blues raised their game. From 20 metres out Carlos Spencer stepped his marker and raced over for the try. With time almost up, the Hurricanes desperately sought to counter-attack but lost the ball on their own line. Craig Dowd grabbed it and smashed over to secure the victory for Auckland.

Professional rugby had started with a bang and was here to stay.

92

BORN TO PLAY SEVENS

29-31 March 1996. Hong Kong Stadium

Christian Cullen takes Hong Kong by storm

Before 1996 Christian Cullen was the best-kept secret in New Zealand rugby. He had played his first game for Horowhenua in 1994 while still at school but though he scored four tries that season, his skills were hidden underneath the massive number of points his team conceded as they were relegated back to the third division. The following season he joined Manawatu, this time scoring 12 tries in 11 games and causing the NZ Rugby Almanack to write of him, "Nineteen year old Christian Cullen scored many brilliant tries with his speed and swerve, leaving would be tacklers grasping at nothing."

That year he also started to play sevens rugby with an impressive showing at the Telecom National Sevens in Palmerston North that brought him to the attention of New Zealand sevens coach Gordon Tietjens. Tietjens probably realised what others would later come to realise – that Christian Cullen was born to play sevens.

In the 15s game he often makes the most amazing runs, slicing through massed opposition ranks with the most subtle of cuts, his own supreme balance throwing out the balance of others, his almost casual-seeming running style outpacing many mere "sprinters." In the sevens game there is simply more space for Christian Cullen to be even more devastating.

The rest of the rugby world was issued a general warning in January 1996 when New Zealand won the Uruguay sevens. Cullen scored 11 tries in five matches and was made player of the tournament.

But the annual weekend extravaganza in Hong Kong is the biggest sevens tournament in world rugby, an unofficial world championship and Olympics rolled into one. All the other teams knew what to expect from last year – a thundering giant called Jonah Lomu.

They soon found out about his little partner in crime, the one who ran through them like they were dummies, the one who caught them embarrassingly with their balance rooted on one foot, the one they were always fingertips away from catching, but never did.

"Nineteen year old Christian Cullen scored many brilliant tries with his speed and swerve, leaving would be tacklers grasping at nothing."

In the first game the Sri Lankans did not know what had hit them. At first they probably felt a relief to be playing against someone who was almost the same height as them, but where was he? Cullen zipped past them to score seven tries in the 75-0 thrashing. After that he just kept scoring them.

He realised later he should have paced himself, but when he looked at the bench and saw great players like Glen Osborne (player of the tournament 1994) sitting on it he felt he had to keep coming up with the goods or he might be dropped. Japan were beaten 77-0, France 28-14, Ireland 49-0 in the quarter-finals, England 42-19 in the semi-finals and Fiji 19-17 in the final.

New Zealand won the title for the third year in a row and broke the tournament record with 290 points. Almost half of them, 136 points, were scored by Cullen with 18 tries and 23 conversions. There was absolutely no doubt as to who was player of the tournament. When the lights went out on the Hong Kong Sevens that year, the star of Christian Cullen was still burning brightly.

93

BLUES FESTIVAL

May 1996. Eden Park, Auckland

Auckland wins the inaugural Super 12

The Auckland Blues had won the first game of the 1996 Super 12. Now could they win the last game and become the inaugural champions? They certainly had the firepower. The team was largely based around the best province in New Zealand,

Auckland, augmented from neighbouring Counties by what one South African writer described as "two exocet missiles on the wings" in the shape of Joeli Vidiri and Jonah Lomu.

Yet the Blues had not had an easy ride to the final. They had lost away to ACT who were supposed to be a team of leftovers, lost away to Transvaal (then coached by Alex Wyllie) and had been completely humiliated by the Queensland Reds 51-13 in Brisbane.

Queensland had won the round robin series but the Natal Sharks had beaten them in Brisbane in the semi-final. So the Sharks had come to Auckland to win and they had some star performers in their line up to do it, such as Andre Joubert, James Small and the man who was to become Springbok captain, Gary Teichmann. Also nagging in the minds of the Blues players was the fact that there had been a Super 10 competition preceding the Super 12 which had run for three years from 1993 – and no New Zealand team ever won it.

A crowd of 45,000 turned up to watch the spectacle in warm sunshine. After blitzing Northern Transvaal at Eden Park in the semi-final the previous week, the

NEW ZEALAND HERALD

Carnival time at Eden Park as the Blues become the first team to win the Super 12.

Blues were ready for more of the same. Within 20 minutes they were ahead with tries to Jonah Lomu and Andrew Blowers.

Natal were not called the Sharks for nothing. They bit back. Henry Honiball made a long break upfield and with Auckland struggling to defend, a try looked certain, but then Lomu got hold of the ball and powered back down the other end of the field. As he was collared by Teichmann, he slipped the pass to Carlos Spencer who scored.

Again Natal fought back, closing the gap to just four points at half time. Then Eroni Clarke got the ball and beat five tacklers, sidestepping fending and swerving his way to the line. Not to be outdone was Charlie Riechelmann who was supposed to be a tight forward, but decided he was really a three-quarter. From 30 metres out he made a swerving run through grasping defenders to top it off with a plunge over the line that would have won a diving medal at the Olympics.

From 30 metres out he made a swerving run through grasping defenders to top it off with a plunge over the line that would have won a diving medal at the Olympics.

His exuberance personified both the manner of the emphatic 45-21 victory and the carnival atmosphere afterwards as Auckland paraded the glittering Super 12 trophy around the stadium to their ecstatic fans. They had proved themselves to be the best provincial team in the Southern Hemisphere and probably the world, scoring over 500 points including 70 tries throughout the tournament

94

THE DEVIL THEY ARE STILL LEARNING ABOUT

15 June 1996. Carisbrook, Dunedin

Cullen, the hammer of the Scots

You have to feel sorry for the poor Scots. In the 5th century it was the Viking invaders who came charging into their villages at dawn swinging great double bladed axes, stealing their stock and their women.

In 1993 at Murrayfield it was golden boy Jeff Wilson who scored three tries against them on his debut in a 51-15 slaughter. In 1995 they were trampled underfoot by Giant Jonah at the World Cup when they All Blacks won 48-30. Then in 1996 a new nemesis came to haunt them.

...the only way to stop him was to build Hadrian's Wall across the middle of Carisbrook.

Though many had marvelled at Christian Cullen's feats at the Hong Kong Sevens, there were those who said he could not hack it in the real world of 15 man test match rugby. All Black coach John Hart was one who thought he could. In Cullen's debut the Samoans decided to try that theory out. Early on winger Brian Lima caught him with a jarring tackle. Then Sam Kaleta decided to tackle him around the throat. Cullen's response was to score three tries. There was great delight in his home town of Paekakariki, though some of the local drinkers felt he could have waited a little longer than the 21st minute to score his first. The landlord of the local tavern had been offering free drinks from the time he first touched the ball until he scored.

But as committed as the Western Samoans were, they were not in the same league as the Scots, the All Black's opponents in the next match. Surely now Cullen would find out what test match rugby was all about.

Scotland had never beaten the All Blacks but they had decided to change that. From the start their forwards thundered into the rucks and mauls, decisively winning them and gaining advantages of territory and possession. The only problem was when you get the ball you are supposed to do something with it. And every time the Scots lost it, the All Blacks, led by Cullen, counter-attacked deep into Scottish territory.

For his first try Cullen cut through the Scots' defence like a knife through butter. Six Scots defenders suddenly realised what it felt like to tackle somebody who was no longer there. They might as well have been trying to tackle Scotch mist.

Cullen obviously believed he was back at Hong Kong playing in the sevens. He bedevilled the Scots with his pace, his anticipation, his footwork and his speed. The Scots seemed to be perpetually off balance as he blitzed them with three more tries, leading them to believe that the only way to stop him was to build Hadrian's Wall across the middle of Carisbrook.

"Where do you find all these players?" asked the Scottish manager Jim Telfer ruefully afterwards. But Cullen's performance was best summed up by the Scots captain Rob Wainwright when he said, "Jonah Lomu is the devil we know, Cullen is the devil we are learning about."

95

AT LAST!

24 August 1996, Loftus Versfeld, Pretoria, South Africa

A historic victory in the heat and dust of South Africa

It seemed that from the beginning of time the All Blacks had been trying to win a series in South Africa. They had come closest at the first attempt in 1928 when the series was drawn. In 1949 the All Blacks were whitewashed 4-0. In 1960, 1970 and 1976 the games were close but South Africa had won.

For 68 years they had been trying and now they had a chance. Coached by the inspirational John Hart and led by the uncompromising Sean Fitzpatrick, a week before they had won the first test of a three-match series. If they could win today in Pretoria they would make history.

South Africa scored first with a forward rush but the All Blacks responded with two superb tries that started in their own half and were both finished off by winger Jeff Wilson.

NEW ZEALAND HERALD

The Incomparables celebrate a great victory. One second later they will collapse to the turf in sheer exhaustion.

Now the All Blacks were seven points clear with just two minutes to go. They needed to stop the South Africans scoring a converted try in order to draw level. They just had to win.

Then a third came from a five-metre scrum. Trickery between Justin Marshall and Walter Little allowed Zinzan Brooke a run at the line with only little Joel Stransky the South African first five-eighth to bump out of the way.

The All Blacks lead 21-8 at half time, but the South Africans came storming back in the second half and touched down for two tries within two minutes of each other.

With 19 minutes still remaining the All Blacks now led by just one point.

First five-eighth Simon Culhane had broken his wrist when South Africa scored their second try. He was replaced by Jon Preston. With his first touch of the ball Preston was called on by Sean Fitzpatrick to take a penalty kick 38 metres out. He calmly slotted it over. Five minutes later he did it again. This time from 50 metres out.

Yet after the South Africans had kicked a penalty at the other end the All Blacks still only led by a slender four points. Step forward Zinzan Brooke. He screamed for the ball from a ruck just outside the South African 22. Zinny swung a mighty boot and drop kicked the ball between the posts.

Now the All Blacks were seven points clear with just two minutes to go. They needed to stop the South Africans scoring a converted try in order to draw level. They just had to win.

But nobody in that South African team wanted to be the first to lose a series at home to the All Blacks. Their own proud tradition rose up and they rallied for one final assault, throwing themselves at the line of black shadows. They came within metres of the line and yet the All Blacks held firm, many collapsing to the turf in exhaustion when the referee's whistle finally blew for a 33-26 victory.

It was the All Black's greatest moment.

96

FOR THE LOVE OF THE GAME

26 October 1996. Blenheim

Division 3 play-off goes down to the wire

While the big boys were getting paid, some of that money trickled down to some of the tiddlers on the lower rung of the rugby ladder. The NZRFU gave a grant to each union, believed to be $22,000. That meant players actually got paid some money and it also helped them to recruit players from outside the region. But $22,000 dollars does not go very far over a season when you have about 22 players in a squad. There was no doubt that in the era of professionalism these players were very much amateurs, playing for the love of the game.

Two of those who came down a notch to keep playing were former All Black lock Chris Tregaskis and former Canterbury skipper Rob Penney, both of whom turned out for Marlborough. Marlborough had a storming season, starting off by hammering last year's beaten finalists Poverty Bay 30-19. North Otago went under the hammer 53-12, conceding 12 tries and then Mid-Canterbury were decimated 51-3.

It was the match in which veteran prop Paul Nicholson made his 100th appearance for his province, but he was easily upstaged by Paul Philips who was playing his 150th game for Marlborough, having started out 16 years before. They finished off the round robin with a 74-26 demolition of East Coast. Marlborough was easily the dominant team in Division 3 and secured home advantage for the promotion play-offs.

After dismissing Mid Canterbury 34-22, they faced a Wanganui team who had been less impressive during the year, beating East Coast by just one point and going down to Marlborough in the round robin match.

Seven thousand loyal fans turned up to cheer on Marlborough in sunny spring weather. The first half opened with Wanganui playing with a strong breeze at their backs. They hoisted a high ball which Marlborough fullback Craig Forsyth failed to gather and Wanganui second five-eighths Jason Hamlin scored the first try of the game in the corner. They went further ahead with a try from a five-metre scrum.

Just before half time Penney made a charge for Marlborough and thrust his way over the Wanganui line to score. Marlborough turned with the wind at their backs in the second half and went on the attack, soon bringing the scores level to 17-17. With 16 minutes to go they were awarded a penalty in front of the posts, but the decision was reversed after the intervention of a touch judge. Wanganui got the ball and ran it up the field for Hamlin to score his second try under the posts.

With minutes remaining Marlborough forced the ball back up to the other end of the field. There followed an amazing succession of five-metre scrums. Such was the pressure that in one of them Wanganui prop Adrian Bull broke his arm.

From another five-metre scrum Penney was able to force his way over the line, but the referee Paul Macfie was unsighted and failed to award the try. Then it seemed first five-eighth Leon MacDonald had grounded the ball over the line for another valid try.

From another five-metre scrum Penney was able to force his way over the line, but the referee Paul Macfie was unsighted and failed to award the try.

But again the referee failed to award it and instead blew his whistle to signal the end of an amazing match which left the Marlborough fans and players shaking their heads with disbelief while the Wanganui players threw their arms into the air in delight at this unexpected victory.

97

VETERAN'S DAY

19 July 1997. Ellis Park, Johannesburg

Old Warrior still runs like a spring chicken

Frank Bunce has played 55 tests for New Zealand (and four for Western Samoa), but he rates the 1997 Tri Nations test in South Africa as the toughest of his career. At the altitude of Ellis Park the pace was so fast at the start, that he says it left him gasping for breath. Nor was he at the top of his fitness, having been injured in the Super 12. But John Hart had put faith in the old warrior and selected him for another season of all Black rugby even though he was pushing 36 years old.

Since South Africa had come back from the rugby wilderness in 1992 they had played the All Blacks 10 times and won just two of those matches. The previous season they had lost their first ever series against the All Blacks in South Africa. This time they came out fired up.

After six minutes they had stormed out to a 10-0 lead. Then Bunce struck.

He got the ball outside the South African 22 with a massed South African defence before him. Yet Bunce seemed to be the only one fully aware of the situation, cutting and stepping through the Boks as if they were statues, then outsprinting the cover over the last few metres. With a final hop to escape a desperate lunge, he dotted down under the posts.

The Bok storm rose again. De Beer dropped and kicked goals, Bennett scored a try and suddenly the Boks had shot out to 23-7. They were on the attack again when Wilson scored an opportunist try, charging down a kick, gathering the ball and then outstripping the cover in a 50-metre sprint to the line. Carlos Spencer scored another try and at half time only four points separated the teams.

After the break De Beer kicked the South Africans ahead. But then Bunce struck again. It was a meant to be a miss move from a lineout. As planned, Bunce broke the South African line inside his own half and looked around for support. There was none, so he kept going. The South Africans parted before him like the Red Sea, covering the pass, waiting for the veteran to unload and sit down for a rest. But Bunce kept going, twice moving the ball to the hand opposite the would-be tackler.

Suddenly he realised he was close to the line. He had succeeded in throwing the South Africans into total confusion. When they saw what was about to happen and two of them threw themselves at Bunce to make the tackle, it was too late. By then Bunce had got a sniff of the tryline and used his strength to score after a thrilling 50-metre sprint. The veteran was blowing a bit as he jogged back.

Spencer's conversion brought the scores level at 26 all. Then de Beer dropped a goal. Spencer kicked a penalty. Then he kicked another to take the lead. De Beer equalised. A Spencer kick took the All Blacks ahead 35-32. De Beer lined up another with only minutes left on the clock.

The All Blacks waited under the posts. De Beer had hardly missed a kick all day. He was looking like the new Naas Botha. But the ball hit the post and there this thrilling, pulsating match ended.

98

WOMEN'S RUGBY WORLD CUP

May 1998. Amsterdam

The gals show the guys how to do it

The biggest problem that the New Zealand woman's rugby team had was deciding on a name. Traditionally, all New Zealand sporting teams are based around the logo and mystique of the original 1905 All Blacks. Therefore the basketball team is known as the Tall Blacks and the netball team as the Silver Ferns. Even the soccer team is known as the All Whites by contrast.

At first the media dubbed the Women's rugby team the "Gal Blacks," but some considered this derogatory. Said NZRFU chief executive David Moffet, "Do you call the women's hockey team the Stick Chicks?" It was months after the World Cup that they finally came up with the name Black Ferns.

Though the "Black Ferns" were strong contenders going into the 1998

...as they went into the semi-final with England they had amassed 256 points including 42 tries in just three games.

Women's Rugby World Cup held in Amsterdam, they knew it would be no pushover. The United States had won the competition in 1991 and England had won it in 1994. Even though the Black Ferns had demolished England 67-0 in a match at Burnham Army camp near Christchurch the year before they expected them to be better for the World Cup.

At the quarter-final stage they disposed of Spain and as they went into the semi-final with England they had amassed 256 points including 42 tries in just three games. They had conceded a total of three penalty goals in the whole competition so far. England were thrust aside in much the same manner they had been in the men's world Cup three years before (though without a drop-kicking number 8 this time). Then New Zealand went on to meet the United States in the final.

In New Zealand such was the interest in the match that an amazing 150,000 people got up in the middle of the night to watch it (more than watched the soccer FA cup final broadcast at the same time). There was vociferous support for the Black Ferns at the match where many ex-pat Kiwis, most of them it seemed male beer drinkers, kept up loud and raucous cheering throughout the match.

The star of the New Zealand team was Waikato winger Vanessa Cootes. Two years before she had been made the New Zealand Woman's Player of the Year mainly for her feat of scoring an incredible 18 tries on a three-test tour through Canada. That's six a game, an average that not even Christian Cullen can match. In style she was more like Jonah Lomu, racing around the outside of her opponents and in build closer to the big black bus Vaiga Tuigamala, though she was undoubtedly better looking than all three of them.

Cootes kept pretty close to her average in the final. She scored a thumping five tries in the resounding 46-12 win over the United States to prove herself the most valuable woman's player in the world and New Zealand the best team in the world.

99

CHAMPION CRUSADERS

30 May 1998. Eden Park, Auckland

One-eyed Cantabrians are All Heart in Super 12 final

It was looking grim for the Canterbury Crusaders as they came into the 1998 Super 12 final against the Auckland Blues. There had never been a Super 12 competition yet that the Auckland Blues had not won. The previous year the Blues had been unbeaten the whole season. They had never lost a home game in three years of competition. Their side was stacked full of current and former All Blacks. *Grim.*

By contrast Canterbury had one current test All Black in Andrew Mehrtens.

NEW ZEALAND HERALD

The Crusaders are a team, Michael Jones is alone and the scoreboard says it all.

Earlier that season they had suffered a disaster when their halfback, All Black captain Justin Marshall, had severed his Achilles tendon while scoring a try. Apart from Todd Blackadder who was also an All Black (but not yet in a test) they were mostly a team of nobodies compared to the "movie stars" of Auckland.

Unlike other franchises, the only first division province the Crusaders could call on was Canterbury, augmented by one or two imports from the draft system, the most notable being the ebullient Norm Berryman who had scored a stunning breakaway try in the semi-final.

In 1996, the first year of the Super 12, the Crusaders had suffered the ignominy of propping up the table. The following year, 1997, they improved to sixth. That year Canterbury the province also won the NPC.

The Crusaders had a stuttering start to the 1998 Super 12 before they hit the winning way. They seemed a good unit, especially in the forwards, but never really dominated sides as the top teams usually do. They seemed lucky to be in the final, having a close call at home in the semi-final when Natal Sharks had dominated a long portion of the game and seemed set to blow them away. They were able to hang on then, but it took a one-eyed Cantabrian to see them upsetting the mighty Blues at Eden Park, the graveyard of so many good teams.

But if rugby is anything it is a game of heart and teamwork. And no team had a heart like the Crusaders. Captain Blackadder wore his on his sleeve, always leading from the front, the rest of the no-name forward pack following behind him. The Crusaders led 3-0 at half time in a defence dominated match, but the Blues took the lead when a snaking run by Stensness led to a try by rookie hooker James Christian. It was another young buck, lock Norm Maxwell, who hit back for the Crusaders when he ran in a try from loose play 25 metres out. That brought the score to 10 all, but the Blues were on fire and turned up the heat. So many times in the past they had upped the tempo to win matches when it really mattered – just before the final whistle.

"I thought we were on toast out there," said Blackadder afterwards. With minutes left on the clock and the scores tied 13-13 the Crusaders managed to push upfield into Blues territory. With the goal posts in sight Andrew Mehrtens was starting to have nightmares about trying a drop goal. Instead the ball was chipped ahead from broken play. Auckland seemed to have it covered, but it was another youngster James Kerr who won the race to touch the ball down for Canterbury.

It was a victory of the new over the old, of heart over experience. While the Auckland team was stacked with stars, the Crusaders had stolen the Super 12 title with the 20-13 win simply because they had played as a team.

100

ALL BLACKS STRIKE GOLD

14 September 1998. Petaling Jaya Stadium Kuala Lumpur, Malaysia

New Zealand cleans up at first Commonwealth Games Sevens

For most sportsmen the Commonwealth Games are a major event in the sporting calendar. They are known as the "Friendly games" where athletes from many

NEW ZEALAND HERALD

Golden boys Cullen, Rush and Lomu celebrate a unique rugby victory.

countries (of the former British Empire) and sports can mix and mingle together.

For the 16th Commonwealth Games in Malaysia it was decided to include rugby as one of the participant sports in the form of sevens. New Zealand has always had a proud sevens record winning the Hong Kong Sevens three years in a row from 1994-6. But in the years since some of their star players had been unable to play due to commitments to the lucrative new Super 12 competition. As the Commonwealth Games were held in September that meant the stars of previous years, including Jonah Lomu and Christian Cullen, were available once again.

In the meantime New Zealand's sevens mastery had been overtaken by the traditional sevens maestros Fiji who had won the sevens World Cup in 1997. From the very outset the Commonwealth Games looked like it would be a two-horse race. The smaller teams were gobbled up in the qualifying round, then Fiji stuttered against Canada in the quarter-finals before pulling through.

New Zealand's moment of tension was in the semi-final where a committed Samoan team attacked from the outset, driving them back deep into their own half. The score seesawed as both teams wrestled for the advantage. Finally New Zealand got ahead, but the match was not over. With seconds remaining New Zealand lost possession near their own line. A Samoan looked as if he was going to score but was dragged down metres short. There was a desperate scramble for the ball before it was kicked out and the match ended.

In the final, the Fijian Prime Minister Rabuka watched from the grandstand but the king on the field was Waisale Severi. He stopped a certain Christian Cullen try with a last ditch tackle that caught Cullen around the ankle and then got up to take the ball and run it back at New Zealand to score. Severi was the smallest player on the field and perhaps the most skilled, but rugby is first and foremost a physical game and both physically and metaphorically New Zealand had a giant of a player in Jonah Lomu.

On one unforgettable run, teeth clenched and nostrils flared he thrust out his hand and bowled over one Fijian of similar height.

On one unforgettable run, teeth clenched and nostrils flared he thrust out his hand and bowled over one Fijian of similar height. As another tall Fijian ran over to make the tackle he suffered the indignity of being bowled into touch. Jonah was on fire.

The match was close but when the final New Zealand try was converted it stretched the lead out to nine points with just minutes remaining. Yet the referee didn't seem to want to end the match. He kept resetting a series of scrums until finally the ball was kicked into touch and the whistle blew.

Never have there been such celebrations on a rugby field. Maybe it was the atmosphere of the Commonwealth Games, maybe it was the difficult season the All Blacks had (losing five matches in a row), but the celebrations were effusive. Stripped to the waist in the sweltering heat, the New Zealand team performed several hakas during the lap of honour, many of them leaping up towards the barriers to "high five" their supporters in the crowd.

Big Jonah, who was undoubtedly the player of the tournament, was reduced to tears. (He later claimed he was wiping away sweat.) These athletes who were now highly paid professionals still played their sport primarily for the challenge and the pleasure. The pleasure was having a gold medal hung around their necks.

BIBLIOGRAPHY

The Best of McLean – Terry McLean, Hodder and Stoughton 1984

Buck, The Wayne Shelford Story, as told to Wynne Gray – Moa Publications 1990

Colin Meads All Black – Alex Veysey, Collins 1974

The Encyclopaedia of New Zealand Rugby – R.H. Chester and N.A.C. McMillan, Moa 1981

The Encyclopaedia of World Rugby – Keith Quinn, Shoal Bay Press 1991

Famous Flankers – Joseph Romanos, Rugby Press 1990

Fields of Glory

Fronting Up, The Sean Fitzpatrick Story – Steven O' Meagher, Moa Beckett 1994

Grizz The Legend – Phil Gifford, Rugby Press 1991

Great Days in New Zealand Rugby, Terry McLean – Collins 1959

Hong Kong, The Sevens – John Blondin, National Publishing Group 1995

Kirwan, Running on Instinct – with Paul Thomas, Moa Beckett 1992

Magic Matches, Great Days of New Zealand Rugby – Graham Hutchins, Moa 1991

Men in Black – R.H. Chester and N.A.C. McMillan

New Zealand Rugby Legends – T.P. McLean, Moa Publications 1987

New Zealand Sporting Disasters, Disappointments and Curiosities – Keith Quinn, Century Hutchinson 1986

100 Great Rugby Characters – Joseph Romanos and Grant Harding (illustrations by Murray Webb) Rugby Press 1991

Our National Game – Ron Palenski, Moa Beckett 1992

Ranfurly Shield Rugby – A.H. Carman, Reed 1967

Real Men Wear Black – Trevor McKewen, Rugby Press 1994

Rugby Almanac – C.A. Akers and N.A.C. McMillan, Hodder Moa Beckett

Rugby Greats Volume 1 – Bob Howitt, Moa Publications 1975

Rugby Greats Volume 2 – Bob Howitt, Moa Publications 1982

Rugby Greats Volume 3 – Bob Howitt, Hodder Moa Beckett Publications 1997

Season of Glory – Bob Howitt, Hodder Moa Beckett 1996

The Visitors – R.H. Chester and N.A.C. McMillan, Moa Publications 1990

Zinny, The Zinzan Brooke Story – with Alex Veysey, Rugby Publishing 1995